PORTFOLIO / PENGUIN

JONY IVE

Leander Kahney has covered Apple for more than a dozen years and has written three popular books about Apple and the culture of its followers, including *Inside Steve's Brain* and *Cult of Mac*. The former news editor for Wired.com, he is currently the editor and publisher of CultofMac.com. He lives in San Francisco.

Jony Ive

The Genius Behind
Apple's Greatest
Products

Leander Kahney

PORTFOLIO / PENGUIN

PORTFOLIO / PENGUIN

Published by the Penguin Group
Penguin Group (USA) LLC
375 Hudson Street
New York, New York 10014

USA | Canada | UK | Ireland | Australia | New Zealand | India | South Africa | China
penguin.com
A Penguin Random House Company

First published in the United States of America by Portfolio / Penguin, a member of Penguin Group
(USA) LLC, 2013
This paperback edition published 2014

Image credits appear on page 306

ISBN 978-1-59184-617-8 (hc.)
ISBN 978-1-59184-706-9 (pbk.)

Printed in the United States of America

Set in ITC Giovanni Std
Designed by Alissa Amell

While the author has made every effort to provide accurate telephone numbers, Internet addresses, and
other contact information at the time of publication, neither the publisher nor the author assumes any
responsibility for errors or for changes that occur after publication. Further, publisher does not have any
control over and does not assume any responsibility for author or third-party Web sites or their content.

To my wife, Traci, and our kids—

Nadine, Milo, Olin and Lyle.

CONTENTS

AUTHOR'S NOTE

The first time I met Jony Ive, he carried my backpack around all night.

Our paths crossed at an early-evening party at Macworld Expo in 2003. As a journeyman reporter hustling for Wired.com, I knew exactly who he was: Jonathan Paul Ive was on the cusp of becoming the world's most famous designer.

I was surprised he was willing to chat with me.

We discovered a shared love of beer and a sense of culture shock, too, both of us being expat Brits living in San Francisco. Together with Jony's wife, Heather, we reminisced about British pubs, the great newspapers and how much we missed British music (electronic house music in particular). After a few pints, though, I leapt up, realizing I was late for an appointment. I hurried off, leaving without my laptop bag.

Well after midnight I ran into Jony again, at a hotel bar across town. With great surprise, I saw he was carrying my backpack, slung over his shoulder.

That the world's most celebrated designer carried a forgetful reporter's bag around all night flabbergasted me. Today, though, I understand that such behavior is characteristic of Jony Ive. He focuses on his team, his collaborators and, most of all, on Apple. For Jony, it's all about the work—but when talking about his work, he replaces I with we.

A few months after our first encounter, I ran into him again at Apple's Worldwide Developers Conference in June 2003. He stood to one side as Steve Jobs introduced the Power Mac G5, a powerful tower computer in a stunning aluminum case. Jony chatted with a couple of officious-looking women from Apple's PR department. After Jobs's speech, I walked over to where Jony stood.

He beamed at me and said, "So nice to see you again."

We shook hands, and he asked in the nicest way, "How are you?"

I was too embarrassed to mention the backpack.

Eventually, I got around to asking, "Can I get a couple of quotes from you?" The PR reps standing by shook their heads in unison—Apple has always been famously secretive—but Jony replied, "Of course."

He led me over to a display model on a nearby pedestal. I just wanted a sound bite, but he launched into a passionate, twenty-minute soliloquy about his latest work. I could barely get a word in edgewise. He couldn't help himself: Design is his passion.

Made from a huge slab of aluminum, the Power Mac G5 looked like a stealth bomber in bare gray metal. The quasi-military aspect suited the times: Those were the days of the megahertz wars, when Apple was pitted against Intel in a race for the fastest chips. Makers marketed computers on raw computing power, and Apple boasted their new machine was the most powerful of all. Yet Jony didn't talk about power.

"This one was really hard," he said. He began telling me how keeping things simple was the overall design philosophy for the machine. "We wanted to get rid of anything other than what was absolutely essential, but you don't see that effort.

"We kept going back to the beginning again and again. Do we need that part? Can we get it to perform the function of the other four parts? It became an exercise to reduce and reduce, but it makes it easier to build and easier for people to work with."

Reduce and simplify? This wasn't typical tech industry happy talk. In releasing new products, companies tended to add more bells and whistles, not take them away, but here Jony was saying the opposite. Not that simplifying was a new approach; it's Design School 101. But it didn't seem like Real World 2003. Only later did I realize that, on that June morning in San Francisco, Jony Ive handed me a gigantic clue to the secret of Apple's innovation, to the underlying philosophy that would enable the company to achieve its breakthroughs and become one of the world's dominant corporations.

Content to stand aside as Steve Jobs sold the public on their collaborations—including the iconic iMac, iPod, iPhone and iPad—Ive's way of thinking and designing has led to immense breakthroughs. As senior vice president of industrial design at Apple, he has become an unequaled force in shaping our information-based society, redefining the ways in which we work, entertain ourselves and communicate with one another.

So how did an English art-school grad with dyslexia become the world's leading technology innovator? In the pages that follow, we'll meet a brilliant but unassuming man, obsessed with design, whose immense and influential insights have, no doubt, altered the pattern of your life.

Jony Ive

CHAPTER 1

School Days

Its hydraulics were so well put together, that it
folded out almost with a sigh. I could see the
incipient talent that was coming out of Jonathan.   —RALPH TABBERER

According to legend, Chingford is the birthplace of sirloin steak. After
a banquet at a local manor house late in the seventeenth century, King
Charles II took such delight in his meal that he is said to have knighted
a large hunk of meat Sir Loin.

Another product of Chingford, Jonathan Paul Ive, entered the world
much later, on February 27, 1967.

Like its latter-day son, Chingford is quiet and unassuming. A well-
to-do bedroom community on the northeast edge of London, the bor-
ough borders the rural county of Essex, just south of Epping Forest.
Chingford votes Conservative, as the constituency of Iain Duncan
Smith, former leader of the Conservative Party, who holds a seat famously occupied by Sir Winston Churchill.

Jony Ive's childhood circumstances were comfortable but modest.
His father, Michael John Ive, was a silversmith, his mother, Pamela
Mary Ive, a psychotherapist. They had a second child, daughter Alison,
two years after their son's birth.

Jony attended Chingford Foundation School, later to be the alma
mater of David Beckham, the famous soccer star (Beckham attended
eight years after Jony). While at school, Jony was diagnosed with the

learning disability dyslexia (a condition he shared with a fellow left-brained colleague, Steve Jobs).

As a young boy, Jony exhibited a curiosity about the workings of things. He became fascinated by how objects were put together, carefully dismantling radios and cassette recorders, intrigued with how they were assembled, how the pieces fit. Though he tried to put the equipment back together again, he didn't always succeed.

"I remember always being interested in made objects," he recalled in a 2003 interview conducted at London's Design Museum. "As a kid, I remember taking apart whatever I could get my hands on. Later, this developed into more of an interest in how they were made, how they worked, their form and material."[1]

Mike Ive encouraged his son's interest, constantly engaging the youngster in conversations about design. Although Jony didn't always see the larger context implied by his playthings ("The fact they had been designed was not obvious or even interesting to me initially," he told the London crowd in 2003), his father nurtured an engagement with design throughout Jony's childhood.

Chip Off the Old Block

Mike Ive's influence reached well beyond the precocious child in his own household. For many years, he worked as a silversmith and teacher in Essex. Described by one colleague as a "gentle giant," he was well liked and much admired for his workmanship.[2]

His skill at making things led to his initial decision to teach handicraft as a career, but a later rise in the educational hierarchy afforded him wider influence. Mike was among the distinguished teachers plucked from daily teaching by the Education Ministry and given the grand title of Her Majesty's Inspector. He assumed responsibility for monitoring the quality of teaching at schools in his district, focusing specifically on design and technology.

At the time, British schools were trying to improve vocational education. A widening chasm lay between academic subjects and hands-on subjects like design, and the latter classes in woodwork, metalwork and cooking—in effect, shop-class subjects—were accorded low status and limited resources. Worse yet, with no accepted standards of teaching, as one former teacher put it, the schools "were able practically to teach what they wanted."[3]

Mike Ive took what came to be called design technology (DT) to a new level, establishing a place for the discipline as a part of the core curriculum in UK schools.[4] In the forward-looking design and technology curriculum Mike helped devise, the emphasis shifted from shop skills to an integrated course that mixed academics with making things.

"He was way ahead of his time as an educator," said Ralph Tabberer, a former colleague and schoolteacher, who would become the director general of schools in Prime Minister Tony Blair's government in the new century. Mike helped write the mandatory curriculum that became the blueprint for all UK schools, as England and Wales became the first countries in the world to make design technology education available for all children between the ages of five and sixteen.

"Under his influence, DT went from being a marginal subject to something that occupied seven to ten percent of students' time at school," said Tabberer. Another of Mike Ive's former colleagues, Malcolm Moss, characterized Mike's contribution to the teaching of design technology: "Mike gained a reputation for being a compelling advocate for DT."[5] In practice, that meant Mike helped transform what was basically a goof-off class into a design tutorial and, in doing so, laid the groundwork for a generation of gifted British designers. His son would be among them.

Tabberer remembers Mike Ive talking about Jony's progress in school and his growing passion for design. But Mike wasn't a pushy stage dad, trying to turn his son into a prodigy like the father of tennis stars Venus and Serena Williams. "Mike's influence on his son's talent was purely nurturing," said Tabberer. "He was constantly talking to Jonathan about design. If they were walking down the street together, Mike might point out different types of street lamps in various locations and ask Jonathan why he thought they were different: how the light would fall and what weather conditions might affect the choice of their designs. They were constantly keeping up a conversation about the built environment and what made-objects were all around them . . . and how they could be made better."[6]

"Mike was a person who had a quiet strength about him and was relentlessly good at his job," added Tabberer. "He was a very gentle character, very knowledgeable, very generous and courteous. He was a classic English gentleman." These traits, of course, have also been ascribed to Jony.

The Move North

Before Jony turned twelve, the family moved to Stafford, a medium-sized town several hundred miles north in England's West Midlands. Sandwiched between the larger industrial city of Wolverhampton to the south and Stoke-on-Trent to the north, Stafford is a pretty place, its streetscapes lined with ancient buildings. On the edge of town the craggy ruins of Stafford Castle, originally established by the Norman conquerors of Britain in the eleventh century, stand guard over the city.

In the early eighties, Jony enrolled at Walton High School, a large, state-run school on the fringes of Stafford. Along with other local kids, he studied the usual grade-school subjects, and seemed to adapt easily to his new hometown. Fellow students remember a slightly chubby, dark-haired, modest teenager. He was popular, with a wide circle of friends, and took part in a number of extracurricular activities around the school. "He was a determined character—he settled in straight away," said retired teacher John Haddon, who taught German at the school.[7]

Although Walton had a computer lab stocked with early computers from the era (Acorns, BBC Micros and one of Clive Sinclair's famous ZX Spectrums), Jony never felt at home there, perhaps because of his dyslexia. The computers of the era had to be programmed, keystroke by keystroke, on a blinking cursor command line.[8]

A local church organization, the Wildwood Christian Fellowship, a nondenominational evangelical congregation that met at a local community center, offered Jony a creative outlet along with other musicians he met at Wildwood. "He was drummer in a band called White Raven," recalled Chris Kimberley, who attended Walton High School at the same time. "The other band members were much older than him. . . . They used to play mellow rock in church halls."[9]

Drawing and design offered another necessary relief from academic subjects as, early on, Jony showed abilities as a skilled draftsman and design technician. His relationship with his father continued to be a source of inspiration. "My father was a very good craftsman," Jony remembered as an adult. "He made furniture, he made silverware and he had an incredible gift in terms of how you can make something yourself."[10]

At Christmas, Mike Ive would treat his son to a very personal present: unfettered access to his workshop. With no one else around, Jony could do anything he wanted with his father's support. "His Christmas gift to me would be one day of his time in his college workshop, during the Christmas break when no one else was there, helping me make whatever I dreamed up."[11] The only condition was that Jony had to draw by hand what they planned to make. "I always understood the beauty of things made by hand," Jony told Steve Jobs's biographer, Walter Isaacson. "I came to realize that what was really important was the care that was put into it. What I really despise is when I sense some carelessness in a product."

Mike Ive also took Jony on tours around London design studios and design schools. One formative moment occurred on a visit to a car design studio in London. "At that moment, I realized that making sculpture on an industrial scale would be an interesting thing to do with my life," Jony later said.[12] By the age of thirteen, Jony knew he wanted "to draw and make stuff," but hadn't quite pinned down exactly what he wanted to do. He contemplated designing everything—from cars to products, from furniture to jewelry and even boats.

Mike Ive's influence on his son's design development may not be quantifiable, but it's irrefutable. He was a strong advocate of teaching empirically (making and testing)[13] and of intuitive designing ("get on and make it, refine as you go").[14] In his slide presentations, the older Ive described the act of "drawing and sketching, talking and discussing" as critical in the creative process and advocated risk taking and a conscious acceptance of the notion that designers may not "know it all." He encouraged design teachers to manage the learning process by telling "the design story." He thought it essential for youngsters to develop tenacity "so there's never an idle moment." All of these elements would manifest themselves in his son's process of developing the iMac and iPhone for Apple.

Jony drove himself to school each day, arriving at Walton behind the wheel of a tiny Fiat 500 that he called Mabel. In early 1980s Britain, many post-punk and Goth teenagers wore black clothes, and Jony was no exception. His long black hair, teased into spikes several inches high, made him look like Robert Smith from the popular band The Cure—though without Smith's heavy eyeliner. Jony's hair spiked so high that it would have been flattened by the roof of his Fiat, so he opened the sunroof. Teachers remember the bright orange Fiat entering the school yard with a mop of spiky black hair poking out the top.

In those days—as now—cars were important to Jony. He and his dad were restoring another car, a vintage "frogeye" Austin-Healey Sprite with spherical headlamps that seemed to rise out its hood like a couple of wide-open eyes. While the look was unusual, giving the small, two-seater sports car an approachable, anthropomorphic appearance, the design was intriguing, too, as the Sprite had a semi-monocoque body, meaning the car's external skin was structural.

At school, Jony's skills as a designer were beginning to emerge. A school friend and fellow design student, Jeremy Dunn, remembered a clever clock that Jony produced. Matte black, with black hands and no numbers, the design allowed the timepiece to be mounted in any orientation. Though made of wood, the clock's black finish was so flawless his friends couldn't tell what it was made from.[15]

With the possibility of university studies looming, Jony began to prepare for A-levels, the standardized qualification tests for university admission in the United Kingdom. His primary emphasis would be design technology, then a two-year combined course. In the first year, students explored the character and capabilities of different materials, ranging from wood and metal to plastics and fabrics, almost any material. The idea was to give students the opportunity to develop ideas and learn practical skills before the second year, which was more academic, and centered on a major project.

"It was very hands on," recalled Craig Mounsey, a designer who took the course at the same time as Jony. "We were being taught execution skills and at the same time design process skills."[16]

Jony's work was exceptional and his drawing excellent. His teachers recalled that they had not seen his standard in another student of his age before; even at age seventeen, his designs were often production ready. "His graphics were brilliant," said Dave Whiting, a faculty member who taught Jony design and technology for several years. "He used to draw initial designs on brown craft paper with white and black pens, which was a really effective and new way to do that. He had a different way of presenting ideas. His ideas were novel, innovative, fresh."[17]

"Jony was so good," Whiting added, "we learned a lot from him, through looking at his work."

Not only was Jony skilled at the crafts side, he was exceptionally good at communicating his ideas. "He did things that other people weren't doing," said Whiting. "When you are a designer, you have to be able to convey your ideas to people who are not designers; perhaps they are financing you or going to do the production, and you have to be able to turn them on to the product and its feasibility. Jony was able to do that."

His teachers recognized how sophisticated his work was, and some of Jony's drawings and paintings were hung in the head teacher's office. "They depicted parts of churches, arches and details of tumbling-down churches and ruins, which were very accurate pencil sketches, as well as watercolors," said Whiting. When the head teacher's office was redecorated some time during the late 1980s, the sketches disappeared, but people remembered his skills. "I have heard Jony say that he is not good at drawing," Whiting said in an interview, "but that's not true.

"Jony saw, even in the early days, the importance of line and detail in products. For instance, he designed some mobile phones that were slim and detailed, like modern phones, even while he was still a school student." Jony's interest in phones was not just adolescent tinkering. He would continue to design new phones through his later schooling (and, of course, at Apple).

For his second-year project, Jony chose to design an overhead projector (OHP). DT students were required to produce initial ideas, refine them, make presentation drawings and mock-up models and, if possible, build the actual product. The task was more than a theoretical exercise on paper: It was a complete design process, concept to completion.

The project also required market research. Jony knew that OHPs were standard issue at schools and businesses at the time. They sat on teachers' desks, projecting images of transparent slides onto walls and whiteboards. The ubiquitous machines were all big and bulky, but Jony, having researched the OHP market, decided there was an opening for a portable model.

He designed a light OHP that would collapse into a matte black briefcase with lime green fittings. Highly portable, it was very modern looking—and quite unlike the clunky, utilitarian desktop OHPs of the day. When the lid of the case was opened, it revealed a Fresnel lens with a magnifying glass and a light underneath. As with traditional OHPs, transparent film images placed on the screen were then projected via a series of mirrors and a magnifying lens onto the wall.

Ralph Tabberer, a teacher friend of Mike Ive's, recalled being impressed when he saw the portable OHP for the first time. "Its hydraulics were so well put together, that it folded out almost with a sigh. I could see the incipient talent that was coming out of Jonathan."

The teachers at Walton liked Jony's project and decided to enter it, along with those of a few other students, into a national competition. That year, the Young Engineer of the Year Award, sponsored by the British Design Council, was to be judged by the internationally famous architect and interior designer Terence Conran. For the first round, the entrants submitted graphics, drawings and photographs. The most interesting designs were then chosen for the next stage of the competition.

Jony's portable OHP project was among those selected for round two.

Before sending his OHP for the next stage of judging, Jony took it apart for a final clean and polish. When he put it back together, however, he inadvertently inserted the lens backward. As a result, instead of projecting a clear image, the inverted Fresnel lens sprayed light in all directions, rendering the image indistinguishable. As submitted, the device was useless, and the judges rejected Jony's design. Still, his idea was certainly unique: Though he didn't win, a not-dissimilar portable OHP hit the market not long afterward.

A Rare Sponsorship

At sixteen, Jony's talent was already beginning to gain the attention of the design world.

Philip J. Gray, the managing director of London's leading design firm, Roberts Weaver Group, spotted Jony's work at a teachers' conference.

As Her Majesty's chief inspector for design, Mike Ive organized what became an annual conference to promote design in the national curriculum. When Phil Gray arrived to be the event's keynote speaker, he laid eyes on Jony's work for the first time.

In the conference foyer a small exhibition of design work by high school students had been installed. Among the work on display were some pieces by Jony. Gray was drawn immediately to Jony's sketches of toothbrushes. Much later, Gray recalled the "fine lines in pencil and crayon," and "the quality of thinking and analysis" evident in the work of the young design student.

"His work stood out as being very mature for a sixteen- or seventeen-year-old," Gray said. "I remarked what an extraordinary talent. Mike replied, 'That's nice because that was done by my son Jony.'"[18]

A few days later, father and son visited Gray at the Roberts Weaver Group offices in central London. Over lunch, Gray gave the Ives some advice about the best colleges for ID. "I made a few recommendations," Gray recalled. His top recommendation was Newcastle Polytechnic.

During lunch, Mike Ive also asked a cheeky question: Would Gray's company sponsor Jony though college? In return for an annual stipend (about £1,500 for each of four years), Jony would promise to work for the design firm after graduation. Sponsoring was very unusual at the time, but Gray agreed.

"Jony is the only person I sponsored at RWG," Gray said. "We had interns, who came to work with us during summer breaks from university—but Jony was the only student we sponsored. . . . I had no problem persuading the other directors at RWG to sponsor Jony, because he had shown some clear talent."

Although it might appear that Mike Ive was driving his son to pursue a career in design, Gray didn't believe that was the case. He thought Mike was just responding to his son's growing obsession with design. "Mike used his position to be able to rub shoulders with the design elite and he hoped that some of that would rub off on Jony," Gray allowed, adding that Jony "was a very smart engineer. . . . Father and son were both very enthusiastic. A liking for design was just in the family."[19]

In the years that followed, Gray had more opportunities to observe father and son. "They were so alike; shy but very focused and [they] always got things done without fuss," he said. "I never recall raised voices! My memories are mostly of smiles and the pleasure of being with them rather than raucous laughter. Mike's pride was there to see but never spoken about. It's unusual but talent and modesty can go together."

His father's influence was evident in Jony's temperament as well as in their shared love of design. "Mike Ive was a real enthusiast who always loved what he did," said Gray. "He was a really energetic person and desperately keen for his son to succeed. He was simply a caring father who tried to make sure that Jony had all the best opportunities to get on as a designer."

In his years at Walton High School, Jony opted to study not only design technology at an advanced level but also chemistry and physics, which was unusual for an arts-oriented student. When he graduated from Walton High School in 1985, he did so with an A in each of his three A-level exams. The hard work of two years of preparation paid off, as earning three top grades wasn't easy: According to UK government statistics, his results put him in the top 12 percent of students nationwide.[20]

These grades made him eligible to apply for Oxford or Cambridge, the best known of the UK universities. Having been interested in studying to be a car designer, he also considered Central Saint Martins College of Arts and Design in London, one of the world's leading art and design schools. But when he visited, the place didn't seem to be a good fit. Jony found the other students "too weird," as he put it. "They were making 'vroom, vroom' noises as they did their drawings."[21]

With his academic record and evident talents, Jony had choices. In the end, he did as Phil Gray had advised and opted for Newcastle Polytechnic in the north of England. Product design was to be his main thing.

CHAPTER 2

A British Design Education

> There is a notion in Britain of a T-shaped de-
> signer: one with depth of discipline in a single
> area but also a breadth of empathy for other ar-
> eas of design.   —PROFESSOR ALEX MILTON

Renowned for its beer (Newcastle Brown Ale), football team (the Newcastle United Football Club), and horrible weather, Jony's new home
was a vibrant, industrial port city. When he arrived at the city on the
River Tyne, Prime Minister Margaret Thatcher ran the country, and the
mainstays of the city's economy, shipbuilding and coal, were in decline.

Despite the rain and Mrs. Thatcher (she'd been really hard on the
miners), Newcastle, located near England's northeast coast, had a
reputation as a party town. Roughly a sixth of the city's inhabitants were
students and the city center was home to many bars and nightclubs. In
1985, Jony's first year in university, the British music scene was as lively
as ever, especially in the North, where bands like The Smiths and New
Order gained national attention. Within a couple of years, the city's
nightclubs would be host to the rave scene, awash with cheap ecstasy
and thumping to the dynamic electronic dance music that Jony came to
love.

Now known as Northumbria University, Newcastle Polytechnic was
(and still is) regarded as the top college in the United Kingdom for ID.
These days, the design school has about 120 staff and admits about 1,600
students from more than 65 countries.[1] The department, then and now,

is housed in a tall building called Squires Building. "It was a rather brutal big building but was a great place for creativity in general," said David Tonge. "It was shared with fine art, fashion and craft just over the corridor. This was before industrial design had become fashionable."[2]

Each floor of the building is dedicated to a different design discipline: ID, furniture design, fashion, graphic design and animation. The department is well equipped with lots of tools and technology. "The designers are able to use a range of materials—wood, paper, plastic, metal, leather, kevlar, cotton, you name it," said Professor Paul Rodgers, who lectures on design at Northumbria, though he didn't teach Jony. "They have access to all these machines—drilling, sawing, fastening, stitching, etching, burning, you name it. And they receive really good training in those workshops, backed by a technical staff."[3]

The ID department at Newcastle, founded in 1953, gained recognition in the sixties, in part because of its close ties to British industry. According to another alumnus, Craig Mounsey, who completed the course a year before Jony, "Newcastle had the reputation for being the best. . . . They won everything. All of the design teachers at school would parade the work from Newcastle as being the standard."[4] Mounsey himself has gone on to become CEO of CMD, one of Australia's leading design studios.

The quality of the student body was another reason for Newcastle's high standing. According to Mounsey, prospective students had just a one in ten chance of gaining admission to Newcastle Polytechnic. In 1984, 250 applicants vied for just 25 places. "We were effectively the cream at the very top of the new wave of school-curriculum-trained designers," said Mounsey. "It was humbling."

The first year at Newcastle was split between learning practical skills and academic classes, with a focus on design psychology. "The first year is a rapid upskilling program," explained Rodgers.

"Students were taught how to think like a designer. One of the first projects was to design two rooms using nothing but several simple geometric shapes: a sphere, cube, tetrahedron and a cone. We had to create one room which would invite the user in and make them feel like they would never want to leave," recalled Mounsey. "The other had to be intimidating and be a place you would want to leave. Polar opposites." The most important part of the project was a report justifying the student's decisions. "The first year was all about thinking, research and abstract design language," said Mounsey.

Students were also required to master hands-on design skills, an emphasis that has continued to this day with the school's focus on project-based learning. Students at Northumbria traditionally spend a lot of time learning how to make things. They are taught how to sketch and draw; and how to operate drills, lathes and computer-controlled cutting machines. They are also given time and freedom to experiment with some of the materials and resources in the school and develop a really deep understanding of what they can do with materials. Throughout this time, the emphasis is on creating and making.

"It's no nonsense," said Professor Rodgers. "We teach the fundamentals. There's lots of emphasis placed . . . on the manipulation of materials."

Another key part of the program is the requirement that students complete two "placements"—in effect, internships—with outside companies. During the middle two years of the four-year program, all the students work in placements in the second and third years. This academic structure is known as a "sandwich" course.[5] While many technical colleges offer placements, most require just one. Northumbria attracts some of the most talented students in the country because of this double-sandwich course structure. Students have taken placements with Phillips, Kenwood, Puma, Lego, Alpine Electronics and Electrolux, among many others, or were placed with design firms and consultancies, including Seymour Powell, Octo Design and DCA Design International.[6]

The program was the same in Jony's time. "It was unusual," said David Tonge, one of Jony's classmates and a close friend. "[The placements] made you much smarter and wiser when you returned. The cumulative effect of everyone doing this and bringing experiences back is huge. Effectively you leave the course with a year or so experience . . . Of course, it's a big leap over other graduates [from other universities]."

The rigor of the coursework and placements give the graduates an advantage in both craftsmanship and the discipline of ID. According to Professor Rogers, "When you look at a Northumbria project and compare it to another institution in the UK, the attention to detail and the making of the artifact is always very, very strong. The things themselves . . . are made to a very high level of detail."

The contrast to Goldsmiths, the famous arts and humanities college in London, is illuminating. Goldsmiths is well known for fostering a generation of high-profile British artists called collectively the Young British Artists (YBAs), including Damien Hirst and Tracey Emin.[7] The YBAs are famous for stirring up controversy and provoking outrage. Hirst pickled dead sharks in formaldehyde and Emin created an artgallery installation of her unmade bed, which included a used condom.

Based in New Cross in south London, Goldsmiths is big city, intellectual, and artistic to a fault. In comparison, Newcastle is blue collar, brass tacks, and a get-your-hands-dirty-making-things kind of place. "At Goldsmiths, the focus is on the idea, the concept," said a Northumbria professor, who asked not to be named. "Northumbria focuses on the object, the artifact. I think, being rather crude here, the focus of the Northumbria graduate is on the detail, and the manufacture and the craftsmanship of producing the object; and a Goldsmith's student would be much more about interrogating a notional product, from a particular conceptual, contextual point of view. In my crude comparison, the Goldsmiths student thinks a lot about what they are doing, whereas the Northumbria student gets on and does it."

The design education Jony encountered at Newcastle was based on a Germanic approach, according to Professor Penny Sparke, pro vice chancellor at Kingston University and a writer about design. "The German Bauhaus of the 1920s was picked up by British design education in the 1950s," she said. "For example, they had what was called a foundation year in Bauhaus, and British design also has a foundation year. The idea of the foundation year was that students started from scratch; they did not build on the past but started on an empty page.[8]

The minimalist principle that designers should only design what is needed also was derived from the German pedagogic tradition. And Ive's design philosophy seems very conscious of that. Both Ive and Braun came out of the same Bauhaus tradition, as have lots of German companies such as kitchen equipment companies, electronics companies—it is quite established in the technological end of German design. There is a vein of high quality, high technology and minimalism. Ive probably imbibed these influences through his education."

Professor Alex Milton, director of research at Heriot-Watt University in Edinburgh, has described the Germanic influence slightly differently. "British education is far more subversive than Bauhaus ever was—in a good way," he said. Milton said more influential was Jony's exposure to all the different kinds of design at Northumbria, from graphics to fashion. Being educated in a giant building with every other discipline of design would have had an influence on the way he would work in the future with multidisciplinary teams, including at Apple. According to Milton, "He would have interacted with fine artists, fashion designers, graphic designers . . . [T]his is something that all UK design students are subjected to—a very broad design education."[9]

"There is a notion in Britain of a T-shaped designer," Milton said, "one with depth of discipline in a single area but also a breadth of empathy for other areas of design. So the British design school/art school vibe informs how Jony Ive interacts with service design, multimedia aspects, the packaging [and] the publicity."[10]

Culture and history have a place in the mix of art and craft to which Jony Ive was exposed in the 1980s. At the time, the nation transformed itself from a semisocialist state with strong trade unions into a fully capitalist one on Reagan's model. There was a lot of youth revolt. Young Brits embraced punk, which encouraged experimentation, unconventionality and daring. It's possible to read some of that independence into Jony Ive's later approach.

"In America, on the other hand," Milton explained, "designers are very much serving what industry wants. In Britain, there is more of the culture of the garden shed, the home lab, the ad hoc and experimental quality. And Jony Ive interacts in such a way . . . [he] takes big chances, instead of an evolutionary approach to design—and if they had focus-grouped Ive's designs, they wouldn't have been a success."

The schooling would distill his work ethic and focus even more. Jony internalized much from his Newcastle experience, including his habit of making and prototyping. His DT education encouraged risks and even rewarded failure, exposing Jony to a very different model from the usual American design school format, which tended to be more prescriptive and industrially focused. If the education system in America tended to teach students how to be an employee, British design students were more likely to pursue a passion and to build a team around them. If this all sounds familiar, it may be that Jony's education in Northumbria prepared him very well indeed for his later career at Apple.

"Jony actually came to Newcastle somewhat unusually; he missed his first day because he was picking up a design prize, which surprised and somewhat intimidated his fellow students. "The first or second day of college, he wasn't there—he was picking up a design award for his work in high school," recalled Tonge.[11]

In the classroom at Newcastle, Jony also encountered individual styles that influenced him. In his first year he took a sculpting class. The professor was allergic to plaster dust and had to wear a mask and rubber gloves, but taught the class week after week. Jony was impressed by the instructor's dedication, but, even more, by the manner in which the professor treated the student sculptures. He took an almost reverential approach to their creations. He would carefully clear all the dust off the students' sculptures before talking about them—even if the work was terrible.

"There was something about respecting the work," Jony said, "the idea that actually it was important—and if you didn't take the time to do it, why should anybody else?"[12]

Newcastle may be a party town, but Jony's memories of this time are less than fun filled. "In some ways I had a pretty miserable time," he said. "I did nothing other than work."[13]

His lecturers remember him as a diligent, hardworking student. "His attitude to work was incredibly thorough," said Neil Smith, principle lecturer, Design for Industry. "Whatever he did was never quite enough; he was always looking to improve the design. He was exceptionally perceptive and diligent as a student. It was never a case of just going through the motions."[14]

"He Looked Like a Hairbrush!"

In his second year at Newcastle, Jony undertook the first of two semester-long placements with his sponsor, the Roberts Weaver Group in London.

At RWG, Jony met Clive Grinyer, a senior designer. Grinyer, who would become a lifelong friend and have a big influence on Ive's life, has himself had a long and fruitful career, even rising to director of design and innovation at Britain's Design Council.[15]

Grinyer and Jony immediately hit it off, despite the age difference (Jony was eight years younger)—and Jony's weird haircut. He had a shoulder-length mullet with a fringe that was back-combed to stick straight up. "He had a little round face and mad hair sprouting out," Grinyer said. "He looked like a hairbrush!"[16]

Grinyer saw beyond the hair and noticed that Jony immersed himself in all the ongoing projects, despite being the office's most junior intern. "The amusing thing is, looking back, that even though there were eight to ten quite experienced designers there, all the work in the studio was going to this student! So Jony was already famous by the time I joined RWG."[17]

Jony and Grinyer shared a similar sense of humor, and Grinyer liked the young man's quiet confidence, even though Ive initially came across as shy and self-deprecating. "He and I immediately became good friends," said Grinyer. "He was ego-free, which was very rare in the design student world. Most design students had lots of ego and very little talent. Jony was the other way round. When designing, he was clearly in love with what he was doing. He became so fixated on all his tasks."

Grinyer had recently spent a year in San Francisco working at ID Two, the U.S. offshoot of Moggridge Associates, a firm founded by Bill Moggridge, the legendary designer who died in 2012. Another well-spoken and articulate Englishman, Moggridge is credited with designing the first laptop, the GRiD Compass, a now-iconic clamshell design of screen and hinged keyboard.

Jony was fascinated by Grinyer's experiences in the States, and peppered Grinyer with questions about America. "Jony was really interested in California," Grinyer recalled. "He was fascinated by the opportunities and the way of life there. Designers are always very aware of the culture of each client for whom they undertake projects, because designers are either enabled or inhibited by the client's attitudes to manufacturing processes such as tooling and so on. And America represented a lot of possibilities for Jony. In the 1980s, the San Francisco Bay Area was a very attractive destination for European designers."

Jony's imaginative designs led to his rapid rise as the company's golden boy and he was placed on an account for the Japanese market. In the eighties, Japan had been like China today, an emerging economic powerhouse. According to RWG designer Peter Phillips, the firm, then one of the top design firms in London, got into Japan by paying a Japanese marketing company to promote its work. The freelance company was expensive, as it took 40 percent of the firm's fees, but worth the price. RWG soon received commissions for all types of Japanese work.

Jony was instructed to work on a range of leather goods and wallets for Japan's Zebra Co. Ltd., a pen manufacturer based in Tokyo. Typical of his style, Jony made intricate prototype wallets out of paper. "I remember him folding and playing with these beautiful all-white folded-out wallets, all double-sided with the leaves," Peter Phillips recalled. "In the corner he'd cut out the tiny detail that showed the embossing. It was an absolute beauty. The most incredible model I'd ever seen. It was stunning."[18] The wallets were one of Jony's first products in white, a sign of the designer's lifelong commitment to the color.

Phillips laughed that Jony, a teenager, was working on the boss's "pet projects" while he and the other salaried designers slaved away on what he called "the dirty ones."

Jony was soon tasked with a new pet project: to create a line of pens for Zebra. After making countless drawings, Jony came up with an elegant design with a special touch that would earn him an immediate reputation in London design circles. Phil Gray, the RWG design director who had agreed to help pay Jony's way through college, remembered the drawings Jony made for the project.

"He created some wonderful rendering techniques that were totally original," said Gray. "He did some beautiful drawings on film whereby he coated the back of the film with gouache [paint] and then turned the film over and did some very fine line work on the other face, so that there was a translucent effect on the drawing. This effect was absolutely brilliant at conveying the materials he was imagining. When he sketched, he was such a fine draftsman that you could not tell whether he had drawn in freehand or used a radius guide. He was that meticulous."[19]

Jony's pen was to be made of white plastic with rubbery side rivets, like small teeth, for a better grip. Again, the product was white, but what set the pen apart from every other was a nonessential feature.

In working out his design, Jony chose to focus on the pen's "fiddle factor." He observed that people fiddled with their pens all the time, and decided to give the pens' owners something to play with when not writing. He cleverly added a ball-and-clip mechanism to the top of the pen that served no purpose other than to give the owner something to fiddle with. The "fiddle-factor" notion may have seemed trivial to some, but the incorporation of the ball and clip transformed the pen into something special.[20]

"That was a new idea back then, to put something on a pen that was purely there to fiddle with," Grinyer said. "He was really thinking differently. The pen's design was not just about shape, but also there was an emotional side to it. This, believe it or not, was quite jaw-dropping, especially from someone so young."

Jony made a prototype that so delighted his boss, Barrie Weaver, that he ended up playing with it all the time. Other designers at RWG noticed, and people started saying the object had a "Jony-ness" about it, a term that suggested an object possessed a sort of unknowable property that made people want to touch it and play with it.[21] Ive's talent for adding tactile elements to his designs was already emerging as one of the young man's trademarks (many of his subsequent designs at Apple had handles or other elements that encouraged touching). His unusual pen anticipated the kind of allegiance that later Jony-designed products would inspire. The pen "immediately became the owner's prize possession, something you always wanted to play with," Grinyer recalled.[22]

Jony's TX2 pen went into production, something almost unheard-of for an intern's design. It sold in large numbers in Japan for many years and, in the memory of his RWG colleagues, was typical of the young designer's work. According to Grinyer, "His designs were incredibly simple and elegant. They were usually rather surprising but made complete sense once you saw them. You wondered why we had never seen a product like that before."[23]

Back at School

After his placement at RWG, Jony returned north. He resumed studying for his degree but, later that year, he won a prestigious travel bursary (grant) from the Royal Society for the Encouragement of Arts, Manufactures and Commerce, better known as the Royal Society of Arts or RSA.[24]

Established in 1754 in a Covent Garden coffee shop, the RSA is an ancient British charity, one of Britain's oldest and finest institutions for the promotion of social change.[25]

The highly competitive bursaries attract entries from hundreds of students all over the country and each bears the sponsorship of a particular company. The RSA grants are, in effect, a recruiting tool, a means for corporations to find hot student designers. The first year Jony entered the Office and Domestic Equipment bursary challenge, the sponsor was Sony.

His winning entry was one of his major college projects, a futuristic concept for a telephone. The phone was a blue-sky project, an exercise in futuristic design, assigned to get the students engaged in What if? thinking. Newcastle put a heavy emphasis on emerging technology at the time, with technologies like the Sony Walkman altering existing modes of listening to music. Though those early devices look primitive today, such portable technology was beginning to become part of everyone's lives. Every student had to have a Walkman.[26]

Students at Newcastle Polytechnic understood their careers would be defined by technology. "We were the guys who were told we had the job of bringing it into the mainstream," said Jony's fellow student Craig Mounsey. "This really was a core part of the course culture . . . [This is] why the course was so successful. We were encouraged to adopt and explore any emerging technology and integrate it into our designs. Further we were encouraged to speculate about future technology directions and their implications."

In responding to the challenge, Jony designed a phone that was an innovative take on landline devices. This was years before the mobile phone became ubiquitous, and his winning design was for an innovative landline phone. Characteristically, it managed to rethink the standard image of what a phone was expected to look like. At the time, phones had a receiver with a headset attached by a coiled wire, but Jony's resembled a stylized white question mark.

He called it, somewhat pretentiously, The Orator. The all-in-white phone was made from a one-inch-diameter plastic tube. The base contained the mouthpiece; the user was to hold the phone by the stalk or leg of the question mark; the curve of the question rose to the earpiece speaker.[27]

The design may not have been very practical, but it was great design. It won Jony a travel award worth £500, which, for the moment, he put aside. As for the phone, set designers for a Jackie Chan sci-fi movie got wind of it and asked to use it as a prop. Jony declined because he thought his prototype was too delicate for use on a movie set.[28]

The RSA hadn't seen the last of Jony. A year later, he teamed up with his friend David Tonge to enter another student bursary challenge. This time, the business services manufacturer Pitney Bowes was the main sponsor of the competition, and the winner would visit the company's headquarters in Stamford, Connecticut.

In their last year at college, Jony and Tonge each had to complete a major project, largely self-driven, and a dissertation as a requirement for the Design for Industry course. Tonge was designing aluminum office chairs, while Jony was working on a hearing aid–microphone combination for use by hard-of-hearing students in a classroom.[29] The hearing aid would eventually be exhibited at the Young Designers Centre Exhibition 1989 at the Design Centre in Haymarket, London, but for the competition, the two soon-to-be graduates were determined to win and so devised another product altogether.[30]

"We felt we could use both our skills to frankly speaking—win," said Tonge. "At that time I was building two fully working aluminum prototype office chairs for my final project and Jonathan was doing his hearing aids. I guess we felt a large object was somewhere in between these two and our joint skills. And we were ambitious."

Jony and Tonge were strategic about entering the RSA travel competition. They reviewed the different project briefs, effectively request-for-proposal descriptions that specified possible entries, before choosing to design an "intelligent ATM." The futuristic ATM promised to be both an interesting challenge and a good fit for their combined skills.

They figured out how to work together to come up with something winning, aesthetically pleasing and useful. Tonge was delighted with the potential collaboration. "It was a scale of product Jonathan enjoyed, could control and excelled at," said Tonge. "The level of finish was what was always amazing about his work relative to others. Others were and are capable of the conceptual thought and creativity but very few capable of that level of finish. . . . [I]t's still the standout component of his work."

Jony and Tonge combined labors to create a flatscreen ATM machine: clean, unadorned and, in the Jony Ive way, made of white plastic. It won the Pitney Bowes' Walter Wheeler Attachment Award, which included a much larger prize than the previous bursary: £1,500.

Years later, Tonge, who went on to have a successful design career at IDEO and now runs his London design studio, The Division, is still proud of their project and the effort they invested. "We did consider the relationship of the piece [the ATM] to users, disability and the space it was living in. It was a very polished piece of work that—without being arrogant—was visually and in detail way ahead of what most students and many professional designers were doing at the time. Hence, I think the judges were just bowled over."

Jony, too, took great pride in his undergraduate work. For his final-year presentation at university, he refined the telephone he'd submitted for the bursary and, when he was ready for his final-year presentation, he invited his friend from RWG, Clive Grinyer, to come up and see it. Grinyer made the five-hour drive from London to Jony's tiny apartment in the tough Gateshead section of Newcastle. When he arrived, Grinyer was amazed to find the apartment filled with more than a hundred foam model prototypes of Jony's project, his design discipline on display. When most students might build half a dozen models, Jony had built a hundred.[31]

"I'd never seen anything like it: the sheer focus to get it perfect," recalled Grinyer.

Grinyer said the differences from one model to the next were subtle, but the step-by-step evolution betrayed Jony's drive to thoroughly explore his ideas and get it right. Building scores of models and prototypes would become another trademark in his career at Apple. "It was incredible that he had made so many and that each one was subtly different," Grinyer said. "I imagine Charles Darwin would have connected with them. It was like watching a piece of evolution really. Jonathan's desire for perfection meant that every single model had a tiny change and the only way he could understand if it was the right change or not was to make a physical model of it."[32]

Jony also invited his sponsor from RWG, Phil Gray, who also vividly remembered the polished final model of the phone. "It was an exquisite piece of design—very cleverly conceived," said Gray. "It was very logical, beautifully thought out. The model was fantastic. Remember, at that time there were no mobile phones. There was no iconic telephone. Telephones were basically a box on the table with a dial or a keypad and a handset over it. So Jony's design was very radical. And very well presented, in terms of its logic and ergonomics—as well as being eminently simplistic."

The professors at Newcastle Polytechnic shared the admiration for the work: Jony's degree exhibition earned him a first, the UK's highest degree distinction.

He gained the respect of admiration of professionals in the field, earning the status of a respected peer at barely twenty years of age. "His exhibition was extraordinary," said Gray.

Jony had also been the first undergraduate student to win two travel bursaries from the RSA. In retrospect, RSA archivist Melanie Andrews, who has helped administer the RSA awards for decades, made a telling point about an early sign of the prodigal Jony's abilities. "In both these projects," Andrews observed, "he displayed an interest in both the hardware and software design of each, which has been the winning formula for Apple products."[33]

Love at First Mac

Two life-defining relationships were firmly established during Jony's college days. The first, made official in August 1987, was his marriage as a second-year student to childhood sweetheart Heather Pegg. Also the child of a local schools inspector, she had been one year below Jony at Walton though the couple met at Wildwood Christian Fellowship. They married in Stafford and would later have twin boys: Charlie and Harry.

Around the same time, Jony discovered another strong love: Apple.

Throughout his school years, he demonstrated no affinity whatsoever for computers. Convinced he was technically inept, he felt frustrated because computers were clearly becoming useful tools in many aspects of life, a trend that seemed likely to gain momentum. Then, toward the end of his time at college, Jony met the Mac.

From the first, Jony was astounded at how much easier to use the Mac was than anything else he had tried. The care the machine's designers took to shape the whole user experience struck him; he felt an immediate connection to the machine and, more important, to the soul of the enterprise. It was the first time he felt the humanity of a product. "It was such a dramatic moment and I remember it so clearly," he said. "There was a real sense of the people who made it."[34]

"I started to learn more about Apple, how it had been founded, its values and its structure," Jony later said. "The more I learned about this cheeky, almost rebellious company, the more it appealed to me, as it unapologetically pointed to an alternative in a complacent and creatively bankrupt industry. Apple stood for something and had a reason for being that wasn't just about making money."[35]

CHAPTER 3

Life in London

> Jony was interested in getting things right and
> fit for purpose. He was completely interested in
> humanizing technology. —PETER PHILLIPS

Summer 1989 saw the departure of Jony Ive, together with David Tonge, for America. Freshly graduated from Newcastle Polytechnic, their RSA prize money in their pockets, the two were booked to spend eight weeks at Pitney Bowes in Connecticut.

If Jony expected to be impressed by what he saw at the company's headquarters in Stamford, about forty miles northeast of Manhattan, he was disappointed. "He did not find it very interesting," Grinyer remembered with a laugh. Jony was much more excited about traveling to San Francisco and touring some of the up-and-coming design studios in the Bay Area.

When their stint at Pitney Bowes was finished, Jony and Tonge split up. Tonge traveled to the offices of Herman Miller, Knoll and a few other firms in the office furniture business, and Jony hopped a flight to California to make the rounds in Silicon Valley. He hired a car in San Francisco and drove down the Peninsula to visit a couple of studios, at one point going to ID Two (now IDEO), where Grinyer had worked, and then Lunar Design in downtown San Jose, which was run by Robert Brunner, a fast-rising design star. He and Brunner established an almost immediate connection.

Brunner was born in 1958 and grew up in San Jose in Silicon Valley, the child of a mechanical engineer father and artist mother. His father, Russ, a longtime IBM-er, invented much of the guts of the first hard drive.[1] Until he reached college, Brunner had no idea there was such a thing as product design. He was on his way to join the art department at San Jose State University when he fortuitously passed a display of models and renderings by the design department.

"I decided there and then that's what I wanted to do," he recalled happily.

While pursuing a degree in ID at San Jose State, he interned at what was then the biggest and fastest-growing design agency in Silicon Valley, GVO Inc. After graduating, in 1981, Brunner joined the firm but grew unhappy, feeling the company had little ambition or vision.

"There was no editorial style at GVO," he said. "They just wanted you to crank out the renderings and keep the clients happy."[2]

In 1984, he tried another tack, teaming up with a couple of other GVO employees, Jeff Smith and Gerard Furbershaw, and another designer, Peter Lowe. The four pooled their money—about $5,000—and leased space in a former helicopter factory. They rented a photocopier and shared a single Apple IIc computer. They named their new firm Lunar Design, a moniker Brunner had been using for his moonlighting work while at GVO.

The timing was perfect. In the mid-eighties, Silicon Valley was just starting to get into consumer products, resulting in a high demand for design agencies like Lunar. GVO also came to the game with a difference—most of the firms in the Valley were run by engineers who had little expertise in design.

"It wasn't like we had a crystal ball or anything," said Brunner, "but it turned out we had very good timing. It was at the launch of the golden age of Silicon Valley. We got started when Frog came over, and ID2 and Matrix came over, and David Kelley, which became IDEO. All that was happening when we started out. It was an amazing time to be working and starting your business in Silicon Valley."[3]

By 1989, Lunar boasted a prestigious roster of clients and was flying high. The clients included Apple, which had Brunner working on several special projects, including an attempt to design a successor to Steve Jobs's original Macintosh, now dated after four years on the market without major changes. (The project, code-named Jaguar, eventually morphed into the PowerPC platform).

When Jony came to visit Brunner, he showed him the tubular phone concept he'd built for his final-year project at Newcastle.

The model wasn't just a mock-up of the phone's shape, like most student projects. It also included all the internal components, and Jony had even worked out how it would be manufactured. "I was really impressed by it," Brunner said. "The design was definitely pushing things a bit in terms of being a usable device," said Brunner. "But what really blew me away was when he disassembled the models . . . [with] all the components inside. I'd never seen a student take a beautiful piece of work and then have it fully figured out. That was pretty incredible." Jony had even worked out the thickness of the parts and how they would be manufactured in an injection molder.

Brunner said that not only were Jony's designs the best he'd seen from a student but that they also rivaled some of the best design work being produced in Silicon Valley at the time. "It was amazing for someone just out of school, so young and who had not had a job yet, to [show] not just the natural ability but the interest in how things worked," said Brunner. "Most students coming out of school are primarily interested in form and image and there's a few that are interested in how things work, but very few that come up with something provocative and amazing—and figure out how to make it work.

"As an industrial designer, you have to take that great idea and get it out into the world, and get it out intact. You're not really practicing your craft if you are just developing a beautiful form and leaving it at that."[4]

Brunner was so impressed that he talked to Jony about the possibility of working at Lunar. It wasn't a formal job offer, Brunner said, but more along the lines of We think you're great, why don't you come and work with us? In any case, Jony said no thanks, that he had promised to return to London to work for RWG, which had supported him through college. It wasn't the only such conversation Jony had on his California trip, as several other companies tried to lure the promising new graduate.

In the years to come, Brunner would prove an important connection for Jony. A few months after Jony's visit, Brunner was recruited by Apple. There he set up the company's first internal design studio, setting the stage for the work that catapulted Apple to the top of the design world. With this, Brunner tried a second time to recruit Jony.

On his return to the United Kingdom, Jony submitted a travel report to the RSA.[5] In it, Jony noted that visiting San Francisco had been the highlight of the trip: "I immediately fell in love with San Francisco and desperately hope that I can return there sometime in the future," he wrote.

Roberts Weaver Group

True to his promise to return to the Roberts Weaver Group, Jony, along with wife Heather, moved from Newcastle to London. His decision to join Roberts Weaver came as something of a surprise to his new boss, Phil Gray, who knew of his other job offers. "He was already being recognized as a very talented 'Young Designer,'" said Gray. "As an honorable person, he accepted our offer—even though he had plenty of other offers at the time."

Joining RWG was more than a consolation prize; Roberts Weaver was one of the top design firms in Britain. Jony joined a talented team and quickly made friends, establishing some relationships that survive today. His friend Grinyer had moved on at that point, quitting to join another design firm near Cambridge, but RWG won several design awards in the late eighties.

Like a lot of consultancies of its kind, the company worked on a wide portfolio of projects, from consumer goods to high-tech products, working for international clients in the United States, Europe, Japan and South Korea. Major clients included Applied Materials, Zebra and Qualcast, a lawn mower manufacturer. The management and production structure at RWG, typical of firms at the time, comprised three different groups working together: product design, interior design and a workshop. Jony was assigned to the product team.

His coworkers comprised twenty designers, engineers and graphic designers working in an open studio. Jony and his colleagues had quick access to a workshop, which was directly below. It was a fully equipped model-making facility, with five on-staff model makers. The interior design team had twenty-three designers, architects and computer specialists.[6]

RWG took on two basic types of product design projects, according to partner Barrie Weaver. One was a full design and development process, typically for clients in the United Kingdom. Such projects involved developing the product concept, producing finished working models, undertaking much of the engineering development and overseeing tooling—in short, taking the project through to production. The second kind of project was more limited, focusing on the generation of new ideas or products, usually for foreign clients, most of them in Japan and Korea. In most such cases, the client company had its own in-house design team but was looking for fresh concepts or a different approach from the outside.

"It is important to understand that our projects were done on very tight time frames and fees," Weaver said of this time at RWG. "If we did not undertake the project efficiently then the business would lose money. This therefore makes decision making prompt and restricts time to be spent on analysis, research, ethnography, societal opportunities, etc."

In his new workplace, Jony was as productive as he had been at Newcastle. "Some designers believe the more research you do, the better the solution," said Weaver. "I personally believe in common sense and intuition. Jonathan's strength was that he quickly grasped the essentials of a challenge, producing intuitive solutions, which were elegant, viable and had a sense of detail rare in one so young."

He gained the confidence of his coworkers as an enthusiastic, hardworking team player. "He was a quiet character, with a lovely sense of humor," Gray recalled. "He was not by any means a loud person in the studio. He was very productive and got on with his tasks. He worked incredibly hard and was very diligent. His productivity was amazing, with consummate quality. He really was prolific. He often would produce half a dozen great ideas in a very short space of time and was not only able to talk about them but could articulate them through some very good draftsmanship."[7]

While Jony's style seemed to suit RWG, the nature of doing business as a consultancy didn't. As an outside firm, RWG often had to make concessions to its clients, something that would soon begin to drive Jony crazy. "It is important to understand that there are other aspects which come into play when you are a consultant," Weaver explained. "The client has the final say—they are paying!"

On the other hand, Weaver clearly understood his designer's frustration. "Sadly, clients' marketing teams often have what I would call dubious taste and force changes upon the design. In consequence you end up with some projects of which you are proud and others which are a compromise."

As part of the design pool, Jony collaborated with the other designers. He worked on outdoor garden lighting and lawn mowers for the UK manufacturer Qualcast. He created several conceptual designs for industrial power drills for another British company, Kango.

His confidence grew rapidly and, after just a few weeks at RWG, he asked Gray for a substantial raise: He was talented and felt he deserved it. But he was also young, just out of school, and Gray had to coach the young upstart on the reality of raises.

"I had to balance the interests of the business," said Gray. "I had to have a very difficult discussion. I had to explain he was on a journey, on a career path. There were others around. Everyone had various strengths and weaknesses. We had to balance the books in terms of making sure everyone got a fair opportunity. That fell upon me to do that. It wasn't a pleasant experience because one doesn't like to disappoint people. But we had a rational discussion. He went away. I think he felt he didn't get the best end of the deal. On the other hand, he didn't sulk. He just carried on."

His talents, however, presented other challenges to his managers at RWG. In 1989, Jony's classroom hearing aid project had been featured in the high-profile Young Designers Exhibition, run by the UK Design Council. The futuristic work came to the attention of an executive at Ideal Standard, a giant in bathroom and toilet fixtures in the United Kingdom. The sales director at Ideal Standard was so impressed with Jony's work that he approached Roberts Weaver and asked that Jony be assigned to work on a particular design project for the company. Roberts Weaver felt obliged to decline the request.[8]

"We had a studio of twelve designers and there was no way we could send a fresh graduate such as Jony to go and work with one of our clients," said Gray. "So we responded to Ideal Standard that they could give us a design brief, which we would fulfill, but it was our decision as to which of our designers we would assign to do the work. When he heard this, he walked away! He specifically wanted Jony to do the project."

In time, the Ideal Standard executive would reemerge on Jony's horizon. But in 1989, the savings and loan banking crisis took a toll on RWG's business. The company's interior design group had been getting numerous commissions to design dealing rooms for banks in the United Kingdom, Spain and Australia. But as the financial crisis spread around the world, the banks canceled. "With the financial crisis, banks terminated projects leaving our designers without work," recalled Weaver. "At the same time, the lack of credit meant UK manufacturers pulled back on their new product development programs."[9]

RWG had to close down the interior design operation in London. Weaver's business partner, Jos Roberts, left, moving to Australia, and the product design team was restructured. As part of the restructuring, Weaver drew up new contracts for all the designers.

None of the designers would sign the new contract—except Jony. And he only signed it because the new contract invalidated his old contract, which had married him to RWG because they had sponsored him through college. As a result of the freshly opened legal loophole, his obligation ended. He quit RWG, the first phase of his professional life at an end.

Tangerine Dreams

Jony went to see his friend Clive Grinyer. Together with another London designer, Martin Darbyshire, Grinyer had cofounded Tangerine Design a year earlier.

The partners were old friends. Grinyer and Darbyshire had met as students at Central Saint Martins in London, and subsequently worked together at the London studio of Bill Moggridge Associates. Grinyer had left to join RWG, where he met Jony, then took a job at Science Park in Cambridge, the UK's version of Silicon Valley. While working in Science Park, Grinyer was approached by the Commtel phone company to design some new phones. Commtel wanted him as a salaried employee, but Grinyer persuaded them to let him take the job as a freelancer. With the £20,000 from Commtel, he set up shop with Darbyshire.

"When I had the opportunity to start up a design consultancy, I asked Martin to go for a curry"—that is, they went out to an Indian restaurant— "and he immediately decided that he wanted to join me. We seem to be glued together through life!"[10] They set up shop in a front bedroom of Darbyshire's middle-class home in Finsbury Park, in north London. Grinyer bought office supplies, including a Macintosh and a laser printer, with the money from Commtel.[11]

Initially they called themselves Landmark, after the Landmark Trust rental houses Darbyshire liked to stay in during his family vacations. They thought the name sounded properly grand, but they were promptly and aggressively sued by a Dutch company already using the name. "We tried to get them to give us a load of money to rebrand, which they didn't do," said Grinyer ruefully. After both sides walked away, the lawsuit died and the partners brainstormed a new name. After agreeing on the name Orange they found it, too, was already claimed, in this case by a group of designers in Denmark.

It was Christmastime, and someone saw tangerines lying around. The name was abstract enough to mean lots of things, which the designers wanted. It also reminded them of Tangerine Dream, an early experimental electronic group that Grinyer liked.

"Looking back, it was the best thing that could have happened because Tangerine is a much better name," said Darbyshire. "It was easy to remember, was understandable in major European languages, and the color was a positive symbol in Asia, a key target market."[12]

In the eighties, design partnerships like Tangerine were uncommon; freelance designers tended to set up shop alone. "In the late eighties, design graduates would have gone into what is known as an 'industry of one,'" explained Professor Alex Milton. "It was an industry of designer–makers, or design art." But Grinyer harbored larger ambitions.

"A partnership felt more like a real business," said Grinyer. "And Martin and I were always interestingly dissimilar—and complementary."

Based in Darbyshire's house, Grinyer continued to work the contacts he had made working in Cambridge. They designed television accessories and hi-fi components and, thanks to some earlier work, were invited to Detroit to give a talk about in-car entertainment.

"I was also writing articles for design magazines," said Grinyer. "Our reputation was building."

Grinyer and Darbyshire made ballsy decisions about promoting themselves. They cleverly made themselves look bigger than they were: As well as writing for design magazines, they took out ads in the same magazines touting their work. The ads got attention, conveying a sense that Tangerine was winning big contracts.

Grinyer and Darbyshire also started teaching at Saint Martins one or two days per week, which helped spread the name of their fledgling company (they taught several designers who went on to become famous, including Sam Hecht and Oliver King). They produced promotional brochures, too, in which they described their work as "products for people."[13]

They described Tangerine's focus as being on end users, who tended to be ignored by other design firms. "No one was speaking about end users," said Darbyshire. "It was all about 'how to deliver things reliably,' not about 'what should it or could it be?' This was the fundamental thing behind Tangerine that I believe we all bought into and worked hard to develop."

The combined marketing strategies worked. "My aim was for us to be among the three most naturally sought-after consultancies to provide advice on product design, along with IDEO and Seymourpowell," said Grinyer. By 1990, they had enough steady work to move out of Darbyshire's front bedroom to a real office in Hoxton, in the East End of London, in a converted warehouse. It was just half of one floor, rented to them by a female architect they knew; the timing was right. "My wife was also about to give birth to our first son, so we needed our big bedroom back," said Darbyshire.

The studio was a classic postindustrial loft, comprising a big, long room, with raw plaster walls and rough wooden floors. The designers set the decorating tone with some Philippe Starck chairs and IKEA desks and shelves. The Hoxton neighborhood today is a trendy area of central London, but two decades ago, tougher times had left a lot of abandoned and derelict light-industrial buildings. Hoxton was also home to lunchtime strip clubs—more like strip pubs, this being London—that catered to workers in The City, London's nearby financial district. Grinyer's car was broken into all the time; his radio got swiped, his tires cut.[14] The London Apprentice pub at the end of Hoxton Street near the studio was a big gay pub in the area, which regularly had ABBA nights that attracted a lot of guests in silver jumpsuits. It was a lively neighborhood.

Jony arrived at Tangerine as a third partner just after Tangerine moved to Hoxton—although he was twenty-three and barely out of design college, Grinyer knew "there was no question of Jony being a junior." Jony and his wife, Heather, bought a small flat not far away in Blackheath, southeast London.

When Jony joined Tangerine, Grinyer and Darbyshire were happy not just to get his immense design talent but a big client: Jony brought with him Ideal Standard, the UK giant in toilets and bathroom fixtures that asked for him personally back at RWG. But at Tangerine, Jony worked on everything, from power tools to combs, and televisions to toilets. At Tangerine, the designers worked on everything collaboratively.

The work was consistent but not especially challenging or prestigious. Tangerine occasionally got commissions from big corporations like Hitachi or Ford, but most of the work was on small projects for random, obscure businesses. "It was a very competitive time for design firms," explained Northumbria's Professor Rodgers. "Companies tended not to specialize. They did everything. They worked on many things— packaging for shampoo, a new motorcycle, the interior of a train. They had to work on everything."[15]

Many of the smaller firms had very finite budgets and little or no experience in working with design consultants. Typically, they expected to spend just a few thousand pounds while Tangerine, still building its business, needed to bill much larger amounts. Often Tangerine's proposals ran into the tens of thousands of pounds, way above their prospective clients' budgets. As a result they wouldn't get the work.

The partners had little choice but to take the rejections philosophically. "Most work in the UK at that time was heavily about engineering, rather than user research or concept design work, so we were ahead of the market a bit," said Darbyshire. "We had, on one hand, to flex and work with smaller companies designing smaller products all the way to manufacture, at the same time as trying to win business from the Asia and the U.S., and grow."

To attract and keep clients, the Tangerine designers worked to make the studio look busier than it was. They remembered a trick that RWG had used: When executives from a car company came to visit, the firm's designers drove their own cars into the studio and put sheets over them, saying they were for a secret project.[16] The trick worked and RWG had gotten the job. Taking RWG's cue, if a client came to visit their offices in Hoxton, Jony and his Tangerine partners made sure the studio was stacked with all the prototypes and foam models they'd created on earlier projects. When the client left, the models would be put back in storage.[17]

Jony, Grinyer, and Darbyshire designed power tools for Bosche and electronic equipment for Goldstar. The three designers gave their full attention to a simple barber's comb for Brian Drumm, a hairdresser in Scotland. Jony's concept contained a spirit level in the handle so that the barber could hold it in the right position while clipping his clients' locks. It's still sold for cutting single-length bobs and other precision haircuts. The job had a small budget, but, characteristically, the designers gave it their full attention. "Brian Drumm chose Jon's beautiful concept for the hair-cutting comb, but I worked painstakingly to translate it into an engineering design for production," said Darbyshire. It was ultimately worth it: The comb went on to win an award from the highly prestigious German Industrie Forum in 1991, burnishing the firm's reputation.

The Hoxton neighborhood suited Tangerine. Jony and Grinyer joined a local gym (Jony to this day does a lot of gym training). "This was old Hoxton, not what Hoxton is now," said Grinyer. "So in the gym there were guys boxing while Jony and I were trying to get fit on running machines and lifting weights." Jony's old friend from college, Tonge, worked just around the corner and visited often. He remembered the area as "very similar to San Francisco's modern South Park neighborhood, near some tough areas but generally full of high-end professionals and a thriving work and art scene. There were also lots of hardware stores and raw-material suppliers, which made it easier to mine for the young designers in the area," including Ross Lovegrove and Julian Brown.[18] "The late eighties was a good vintage. ID was not yet fashionable so a lot of people were doing it for the right reasons—to make good design, not become stars."

A year after Jony joined Tangerine, the three principals were joined by a fourth team member, Peter Phillips. The men shared an interlinking work experience: A 1982 graduate of Central Saint Martins College with a BA in ID, Phillips met Grinyer at Central, knew Darbyshire when they both worked at IDEO, and had encountered Jony at RWG.

"When I first met [Jony], he was just starting," said Phillips. "My impression of him was that he was a really nice bloke, just one of these delightful gentlemen. He is what he looks like. He does have this very quiet demeanor about him. Very generous, but he wasn't very serious, always had the ability to just laugh about things. But he was bloody good at what he did."[19]

Like Jony, Phillips brought some clients with him, including two electronics giants, Hitachi and LG Electronics. In the middle of the recession, LG especially was a huge win. The Korean giant had set up its first Euro design center in Dublin, Ireland, and wanted a European design firm. "We got into LG early on and all were very enthusiastic," said Phillips. "It was fantastic and allowed us to design some great stuff."

The four designers were equal business partners in the venture. There were disagreements along the way, mostly between old friends Grinyer and Darbyshire, but they got through them. "Sometimes it came down to who could shout loudest but we always came to an amicable arrangement," said Phillips. "It was something we'd brush under the carpet after a few minutes. Jony and I were the diplomats of the group, we used to say." The young business also had to be careful with its finances. "We were sensible people so we never really pushed the finances too much," said Phillips. "If we had a small overdraft or didn't have enough money at the end of each month to pay ourselves, we would take less and do it sensibly."

The Tangerine crew in those days still made lots of detailed sketches and models. Many were made in Jony's parents' garage because of the mess, but others were farmed out. "We had a good model maker that was a stone's throw from the studio who made sure the models were good and beautiful," said Phillips. "Most of UK model making has gone bust now but they were great at the time. They'd be craft-based, and we'd go down there and discuss details."

Technology was becoming a part of the design process, but slowly. One Mac sat in the middle of the room, and, as Phillips reported, "We took turns to use it." Tangerine was typical in this, as computer-aided design (CAD) hadn't yet become essential to most designers' tool kits.

Emerging Style

A voracious reader, Jony's tastes ran to books on design theory, the behaviorist B. F. Skinner and nineteenth-century literature.[20] A museum-goer, too, he and his dad had made many visits over the years to the Victoria and Albert Museum in London, one of the world's leading art and design museums.

He studied the work of Eileen Gray, one of the twentieth century's most influential furniture designers and architects. Modern masters fascinated Jony, among them Michele De Lucchi, a member of Italy's Memphis group, who tried to make high-tech objects easy to understand by making them gentle, humane and a bit friendly.[21]

Grinyer remembered Jony falling in love with furniture maker Jasper Morrison's school of design, which was very architecturally pure, all straight lines with no curvaceous shapes. He was also fascinated by Dieter Rams, the legendary designer at Braun. "We were all inspired by Dieter Rams," said Grinyer. "Rams's design principles were implanted into us at design school—but we were not designing products that looked like Braun's at Tangerine. Jony just liked the simplicity."[22]

All four designers were interested in design philosophy, but Jony especially. Because Grinyer and Darbyshire were teaching, they were interested in trying to articulate what the firm's design philosophy was. Both Grinyer and Darbyshire had done time with Bill Moggridge at ID Two/IDEO; he was a big influence on their style. One of the key lessons was adopting "no strong ideological viewpoint," recalled Grinyer. Another key for Moggridge had been collaboration.

"IDEO had a consensus [system] so that everyone had to agree, and so Grinyer and Darbyshire were quite keen that when they did design work that everyone agreed on it. So they had a lot of reviews together about designs going forth, which was very good because it meant that you were testing yourself all the time," Phillips said, "and that's a really good way of doing it so that you are trying to please your client and at the same time you're trying to push yourself, because you begin to feel yourself that this is something quite exciting."

As for aesthetics, there were influences, but the Tangerine group never clung to a style for its own sake. "It was important for all of us, including Jony, that we were designing things for a reason," said Phillips.

"Jony was interested in getting things right and fit for a purpose. He was completely interested in humanizing technology. What something should be was always the starting point for his designs. He had the ability to remove, or ignore, how any product currently is, or how an engineer might say it must be. He could go back to basics on any product design, or user interface design. And we all shared this similar philosophy at Tangerine. It was not so much due to our formal design education, but more a reaction to seeing the ways that other people were designing."[23]

For Jony, this represented a shift. At college, his projects had showed the beginnings of a design language, or at least a signature style, as most were in white plastic. But at Tangerine, Jony went out of his way not to put a particular stamp on his work. "Unlike most of his generation, Ive did not see design as an occasion to exert his ego or carry out some pre-ordained style or theory," wrote Paul Kunkel, who interviewed Jony in detail for his book, *AppleDesign*, a look at Apple's design department in the 1980s. "Rather, he approached each project in an almost chameleon-like way, adapting himself to the product (rather than the other way around) . . . For this reason, Ive's early works have no 'signature style.' " 24

Then, as now, Jony's aesthetic tended toward minimalism, in reaction perhaps to the mid-1980s tendencies for excess. That had been the height of the "designer decade," when the splashy colors of Culture Club and Kajagoogoo represented good taste. According to Kunkel, Jony avoided styling his products to protect them from dating too quickly. "In an era of rapid change, Ive understood that style has a corrosive effect on design, making a product seem old before its time. By avoiding style, he found that his designs could not only achieve greater longevity, he could focus instead on the kind of authenticity in his work that all designers aspire to, but rarely achieve."

Jony was not alone. Grinyer, Darbyshire and Phillips were minimalists, too, as were a growing number of other design firms. There was a global wave of minimalism, adopted by Tangerine and picked up by other designers, among them Naoto Fukasawa in Japan and Sam Hecht, another Saint Martins graduate, who worked on a lot of design for "no-brand, high-quality" Muji, the consumer and household products manufacturer. "In the contemporary culture of the 1980s, there was the cliché of the over-designed environment, where everything was a riot of color and form," explained Professor Alex Milton. "It was a visual overload. Objects shouted at you.

"[Jony] graduated in this period where there was a lot of over-design. Objects did not impart any of their owner's personality. They were brands. And so designers wanted to become cooler, calmer, more reflective, and return to a sense of functionalism and utilitarianism."

Darbyshire expressed Tangerine's basic thinking this way: "We were trying to make things genuinely better, giving thought to the visual quality, usability and market relevance of all that we designed."

Grinyer contrasted this approach to the work at other agencies, which tried to put their signature on their commissions. "When I was with Bill Moggridge, I saw lots of really good designers who could only design a particular sort of office-based industrial product," he said. "When they tried to apply their same aesthetic to more mass-market, everyday products, they really failed. They came up with oddly techno products. And that puzzled me. I thought that design should be able to speak in different languages according to each specific purpose."

Advances in manufacturing technology allowed Jony and his colleagues to gently push the envelope. "The 1990s were a time when we were beginning to be able to decorate products," Grinyer said. "Their form could be more interesting. It was no longer just about cladding electronics and putting the button in the right place. We could bring in more shapes, exploit the fluidity offered by injection molding plastics. We could create things that were actually beautiful rather than simply functional."

Again, this had a potential downside, which Grinyer witnessed firsthand working at IDEO. "Designers would often come up with merely a shape," he said. "They did not think about the different functionality of, say, a computer screen and a television. I thought that was a mistake. We did not want to make something that was merely a beautiful shape. We wanted our designs to fit into people's homes. And we were very much focused on the user interface of products."

For his part, Jony took an independent view: His priority seemed always to be the creation of objects that were beautiful rather than simply functional. He was constantly questioning how things should be. "He hated ugly, black and tacky electronics," recalled Grinyer. "He hated computers having names like ZX75 and numbers of megabytes.

He hated technology as it was in the 1990s." At a time of big changes in design, Jony looked to find his own way.

Frustrations

Ideal Standard had been rebuffed when it asked to work directly with Jony at RWG, but Tangerine was only too happy to oblige the company. Early in Jony's time at Tangerine, Ideal Standard commissioned a new line of bathroom appliances, including a toilet, bidet and sink, to replace the company's long-running Michelangelo line.

With characteristic thoroughness, Jony, Grinyer and Darbyshire applied themselves to the new bathroom suite. But the work, which had seemed like a lifesaver for the young company, soon turned into a nightmare. Looking back on it, the episode probably planted the seeds of Jony's eventual departure.

As he threw himself into the work for Ideal Standard, Jony bought marine biology books for inspiration and scoured them for influences from nature. "Jony was very fascinated with water, and he looked a lot at water flow," recalled Grinyer. "He took his inspiration for the bowl he designed as almost a Greek religious artifact.

"He talked about the worshipping of water. That water was going to become a scarce resource. That water was something that must be honored. So he made an elliptical bowl, with an incredible architectural pillar. It was beautifully radical."[25]

Jony, Grinyer and Darbyshire presented Ideal Standard with three collections, each named after Ninja Turtles that, in turn, had gotten their names from the Renaissance artists Raphael, Donatello and Leonardo.[26] They went to Jony's parents' garage—the senior Ives by then had moved to rural Somerset—and produced handmade foam models of sinks and toilets. They were "stunningly good," said Grinyer.

The three designers drove to Ideal Standard's headquarters in Hull to pitch the company on the new designs. They set up in a large room so that a product manager could eyeball the models to see if they were fit for presentation. When they were ushered into a meeting with the CEO and a couple of other executives, however, they failed to impress.

The CEO rejected the designs outright in a torrent of criticism. They were too expensive to produce, he said, and they didn't fit into the established design line. The executive worried that the sink's architectural pillar, which Jony had been very proud of, might fall over and crush children.

"The presentation was hard work," remembered Grinyer. "Our ideas for the designs fell flat. . . . The problem was that our designs were very different to Ideal Standard's usual range of products."

To make matters worse, the CEO was wearing a red clown's nose made of foam. It was the UK's Comic Relief Day, a nationwide fundraiser where everyone wears red clown noses to raise money for charity. The rejection seemed like a bad joke indeed.

Grinyer said Jony drove back to London dispirited. "He was dejected and depressed," recalled Grinyer. "He had poured himself into working for people who really didn't care."[27] Darbyshire remembered that day too. "Jony was unhappy that they hadn't completely loved his designs."

Jony's good humor usually allowed him to laugh off such setbacks, but the crashing failure at Ideal Standard wasn't so easy to forget. Even though the dejected Jony would continue the development of the design, the process felt wrong to him. According to Darbyshire, "The problem was that they wanted to 'productionize it' and, in so doing, tore out its heart and soul."

Despite the initial failure with Ideal Standard, Tangerine was beginning to work with more and more big clients. The team, according to Grinyer,

felt as if they were on a "rollercoaster." They continued to cleverly pub-
licize themselves. "We did a lot of speculative pieces whereby I came up
with some concepts and Jony did amazing design works, which we
could then push out to the press, get them beautifully photographed
and create a buzz," said Grinyer. "Within just five years, we went from
a small company operating out of a back room of a house in London to
having big international customers."

Jony didn't enjoy the fact that part of his job involved selling the
firm. "At Tangerine, it was always very important that the four partners
were the creatives," Grinyer explained. "We were perhaps proud of
ourselves and we wanted to carry on designing rather than be figureheads
who took on commissions and passed them on to behind-the-scenes
employees to do the work. However, despite this, as an agency we were
spending up to ninety percent of our time selling our services. Jony was
younger than we were, and wanted to devote all his time to designing
great stuff. He could sometimes feel frustrated."

Jony gradually realized that he wasn't cut out for consulting. He
loved to design, but found the concessions necessary to build the
business difficult, resisting the reality that, in small partnerships, every
designer should also be a salesperson.

"I was pretty naïve," Jony said later. "I hadn't been out of college for
long but I learnt lots by designing a range of different objects: from hair
combs and ceramics, to power tools and televisions. Importantly, I
worked out what I was good at and what I was bad at. It became pretty
clear what I wanted to do. I was really only interested in design. I was
neither interested, nor good at, building a business."[28]

It particularly upset him when his work got ruined by the people he was
working for. His old boss at RWG, Phil Gray, spoke to Jony in 2012: "Jony
told me that the reason he was so frustrated in consultancy was because he
was not able to see projects through to completion," reported Gray. "Clients

would select bits of what he had produced and then instruct him how to put together those bits according to their own ideas. He was not able to execute what he had in mind for any project. He was so far ahead of his time, that clients often just did not get what he was doing."[29]

Grinyer had also seen Jony's frustration with the clients. "Frequently he would make a beautiful design, which when engineered would look only half as good as it should have done."

Brunner Comes Calling

Bob Brunner, whom Jony had met on his California trip several years earlier, paid a visit to Tangerine's studio on Hoxton Street in fall 1991. Having left Lunar Design three years before, Brunner had settled in at Apple, and was now the head of ID. He'd built a killer team of hotshot designers (including several who would later play important roles in developing the iPod, iPhone and iPad).

Brunner was scouting Europe for outside design firms to work with Apple on a secret project called Project Juggernaut. Even though it was officially taboo for a big company like Apple to use outside commissions to recruit talent, Brunner later admitted that was one of his goals.

"I was trying to get Jony," he admitted. "I wanted to get him to work on the project, and I thought it was another way to get him involved with the company."[30]

In 1991, Apple was still riding high. The company had grown from a tiny start-up in Steve Jobs's garage to one of the largest companies in the fast-growing PC industry. Steve Jobs was no longer at Apple, having quit six years earlier, and was now trying very hard to make his new company, NeXT, a success. His other company, Pixar, was also struggling although, four years later, it released its first film, *Toy Story*, which became a blockbuster.

Apple was being run by John Sculley, a former PepsiCo executive whom Steve Jobs had lured to the company with the immortal line: "Do you want to sell sugar water for the rest of your life, or do you want to come with me and change the world?"[31] Sculley's reputation is mixed these days, but at the time, he hadn't yet put a wrong foot forward. Apple was huge, and the computer industry was exploding. The desktop publishing revolution was putting Macs in businesses all over the world. The company had just celebrated its first ever two-billion-dollar revenue quarter. *MacAddict* magazine proclaimed the era was "the first golden age of Macintosh." Windows 1995 was still a few years in the future; no one could yet know that Microsoft's operating system would entirely reshape the PC industry—and almost put Apple out of business.

Flush with cash, Apple was expanding its product lines. Sculley was investing Apple's horde of $2.1 billion in cash in R&D to speed up development of new products. He got a lot of attention for talking up a new line of innovative hyper-portable computers he called "personal digital assistants," a term he coined at a major speech at the Consumer Electronics Show in Las Vegas.[32] Although it would be a couple of years before Sculley's PDA would actually hit the market as the Newton MessagePad, the IDg was hard at work on it.

Brunner's design team was busy not only with the MessagePad but also a new line of PowerBooks. Although the first PowerBook hadn't yet been released, Brunner's design team was working on the second generation. The PowerBook was a revolution: the first "real" laptop in the fledgling PC industry, which had concentrated on desktop machines. However, the first PowerBook was big and heavy: It was more like a battery-powered desktop than a true portable computer and, for the design team, it'd been a nightmare. Brunner and his team had had to simultaneously invent, engineer, design and test the machine under a tight deadline.

Faced with the work of the next generation of products, Brunner was worried that the team was concentrating too much on the here and now, and not spending enough time looking at what might be coming up in the future. It was obvious that mobile was a new frontier, and Brunner wanted to see where it might go.

"As product schedules get tighter and the level of difficulty rises, the first casualty is innovation," Brunner said. "I wanted to see design that was leaning forward . . . design that would predict what was coming rather than reflect what we already know and see."[33]

Attempting to keep the spirit of innovation alive, Brunner had started conducting offline projects—what he called "parallel design investigations."

"The idea was to develop new form factors, new levels of expression and strategies for handling new technology without the pressure of a deadline," he explained. Critically, Brunner wanted to keep this type of investigation "off-line" because it allowed his team to make mistakes, to feel separate enough from the grind of production that the creative juices could percolate. "Because the ideas generated off-line are often our best ideas, parallel design investigations can be extremely valuable," he said. "This information not only enriches our language, it gives you something to point to and say 'this is what we can move towards.'"[34]

That was a factor in his European trip: Brunner especially liked working with outside consultants (after all, he'd been one himself with Lunar Design before moving to Apple). "I decided to hire people from the consulting world so that IDg could function like a consultancy working with the speed and agility of an independent design firm," he said. "In my experience, consultants want to build their portfolio and will compete to do the most interesting work. So I centered my search on the best local consultants and talented people, fresh out of school." This description fitted Jony and Tangerine to a tee. Jony was only three years out of Newcastle and had impressed Brunner with his concept phone.

As soon as he walked into the Hoxton studio, Brunner was excited by what he saw. He first laid eyes on a soda-making machine created by Grinyer for SodaStream, a British company. It had a swing door that opened and closed with a clever latch—the kind of hinge that, Brunner thought, might be adapted to the screen on a portable product. Brunner raved about it: "This is exactly the kind of creative thinking we're looking for."[35]

Brunner had something of his own to show the four London designers. He pulled a prototype PowerBook out of his bag. Phillips was impressed. "I had never seen that and it was incredible," he said. Indeed, with its offset keyboard, centralized pointing device, and forward palm rests, the first PowerBook would set the standard for basic laptop design for the next twenty years, a fact that still surprises everyone. "We hit a homerun with the PowerBook," Brunner said. "It surprised me to death. There were so many flaws with that machine and that design. I thought it was going to be a huge failure. But looking back today, basically all laptops are that design—a recessed keyboard, palm rests, a central pointing device."

Before Brunner's design, laptops had keyboards that were all the way forward and they didn't have pointing devices. Most ran Microsoft's MS-DOS, which relied on a command-line interface, rather than a graphical one like the Mac, and so used four-way cursor keys. There was no need for a pointing device and, as Windows took hold, manufacturers started using clip-on trackballs.

"It's interesting, in retrospect, that from the PowerBook to today's MacBook, it turned out to be almost the perfect design," Brunner said. "No one's been able to improve on it. . . . [W]e didn't realize we already had something that was very difficult to move beyond."

Over the course of a few London meetings, Brunner and Tangerine exchanged ideas. Jony created a prototype mouse as a sort of test. The conversations went well and, as a result, Tangerine was given a contract to consult on Juggernaut.

Jony was both excited and scared. The Apple job was a huge break for Tangerine and, he realized, for him personally too. He later recalled, "I still remember Apple describing this fantastic opportunity and my being so nervous that I would mess it all up."[36]

As Brunner explained, Project Juggernaut was a wide-ranging parallel product investigation. The idea was to explore a suite of mobile products even further off in the future. Brunner and his team felt confident that the new PowerBook and Newton portable would kick off a whole range of mobile products. They began imagining noncomputer products, including digital cameras, personal audio players, small PDAs and bigger pen-based tablets. (These might sound familiar but fulfillment of these dreams wouldn't come for at least another decade—and under the leadership of an entirely new regime.)

They hoped that pen-based digital assistants, digital cameras and laptop computers could be linked together using infrared, radio wave and cellular networks. Brunner wanted the design group to have several mobile products ready in case Apple's upper management suddenly decided the company needed to start making them.

Brunner had approached a couple of other outside design firms in addition to Tangerine, and he had some of Apple's in-house designers working on concepts. "We knew certain things were coming," explained Brunner. "We knew wireless was going to be important, and that image capture was going to be more important. Things were going to get smaller. Batteries were going to get better."[37]

While an Apple team in California worked on several concepts for portable products, the team at Tangerine designed four speculative products: a tablet, a tablet keyboard, and a pair of "transportable" desktop computers. Brunner wanted the products to be convertible; the tablet should convert into a laptop and vice versa. "For some reason, ironically, we thought convertibility would be really important," explained Brunner. "So you could go from a traditional keyboard and mouse mode to a pen-based mode, which is a little bit of a rage today with some subnotebooks."[38] Brunner noted that these ideas, which seemed a bit strange and radical in the early nineties, weren't a million miles from the latest tablets and hybrid laptop/tablets for sale today.

Brunner asked Jony and his Tangerine colleagues to push the boundary of the design but to keep the main elements of Apple's then-current design language (mostly dark gray plastic with some soft bulges). The designs had to be based on real technology so they could conceivably be real products in the near future.

Jony, together with help from Grinyer and Darbyshire, worked on a tablet called the Macintosh Folio. It was a chunky, notebook-sized tablet with a pen-based screen and a huge built-in stand. Made of Apple's then-usual dark gray plastic, it could almost be a predecessor to the iPad, despite being about five times thicker.

Jony worked alone on a special smart keyboard for the tablet called the Folio keyboard. But unlike today's detachable keyboards for modern tablets, the Folio keyboard was conceived as an "intelligent keyboard" because it featured its own CPU, network jacks and a trackpad. In effect it was half a laptop, namely, the keyboard half.

Grinyer and Darbyshire worked on a pair of "transportable" desktops that were half desktop, half laptop. These were transformer-like computers, convertible machines with built-in keyboards and screens that could transform from a desktop into a portable and back again.

One was the SketchPad. Made of light gray plastic, it featured an articulated screen that could be adjusted for height and tilt. It could be folded up into a purse-shaped bundle with a carrying handle for easy transport. (Jony would later revive the idea of a built-in carrying handle with the first iBook.)

The second transportable desktop was called the Macintosh Workspace. It had a built-in, pen-sensitive screen, along with a split keyboard that folded underneath and to the sides when not in use. The Workspace could be folded flat—like a big fat tablet—for transporting, but spread its keyboard like a pair of chunky wings when opened for use.

"I remember one day seeing Jony with this foam model of the tablet on his desk," said Phillips. "He was sitting away from it with his knees up, typing away on this foam keyboard, and saying 'this feels good.'"[39]

"Jon's was the most thought provoking and, as ever, beautifully detailed," recalled Darbyshire of the Folio. "He sweated and sweated to get it right and he did just that. It was stunning."[40]

With remarkable speed, Jony and the other designers on Project Juggernaut had developed about twenty-five models. In a matter of weeks, they presented the work to Brunner and his team and, over the next few months, the concepts were refined to four principal designs.[41]

As the project neared completion, Jony's fears about messing it up almost came true. The fledgling company didn't have its own model shop for making final prototypes, as such shops require special skills, tools and personnel, typically beyond the means of all but the biggest design studios. (Today, even a company as big as Apple uses external shops for finished models.) So the four Tangerine designers took their Juggernaut mock-ups to a local model maker who had done a lot of work with the film and advertising industry.

The model maker was very talented and his models looked "fantastic," said Grinyer. They were perfect for showing off their ideas to clients. However, the models weren't made to last. "When we got the models back, they looked fantastic but would break after you had used them once," said Grinyer. "Apple had this pile of broken models they couldn't do anything with. That was all a bit of a disaster."[42]

Despite the rickety models, Brunner was mightily impressed with Jony's contributions.

"Jony did a really, really sweet tablet," said Brunner. "Really amazing. True to form with Jony, it was very clean, very sophisticated, super attention to detail. But still it had provocative qualities. . . . It was very developed, refined, sophisticated surfaces that were clean and beautiful but they still felt emotional. They weren't dull and boring."[43]

Brunner remembered that Jony's Juggernaut designs stood out because they weren't based on anything that Apple—or any other computer company—had done before. They were utterly original. "They had an emotional maturity that's rare for someone Jonathan's age," Brunner added.[44] Jony was twenty-six at the time.

After six months of work on Project Juggernaut, Jony, Grinyer and Darbyshire were flown to Apple's Cupertino headquarters to make a final presentation. Since Phillips hadn't been involved in the day-to-day work on Juggernaut—he'd been keeping the company afloat working for LG—he stayed behind.

Both Jony and Grinyer liked the vibe at Apple, but Darbyshire found it cliquey. "Apple is a deep culture," he said. "You have to want to belong to it. It's almost to the point of being a cult, and I have to say I find that kinda spooky. There are fantastic sides to the culture as well: a great sense of freedom, constant encouragement to develop and look for something new to perfect, but also a slightly cliquey weird side that I find too claustrophobic. It's almost a religion and I can't deal with that."[45]

After the presentation, as they were packing up to go, Brunner pulled Jony to one side to speak to him privately. He told Jony that if he really wanted to "create something radical," he should come to work for Apple full time.[46]

"It wasn't super overt," said Brunner. "It was more along the lines of me mentioning that the opportunity still exists, and him saying, 'That's pretty interesting, let me think about it.'"[47]

Jony did think about it, and back in London, he agonized over the decision. He had enjoyed working with Apple, but wasn't sure if he wanted to leave both his homeland and his work at Tangerine. He also wasn't sure whether his wife, Heather, would want to move to the States. But Project Juggernaut had opened a new window for Jony.

"Even though I had done a lot of interesting work up to that time, the issues I encountered on Juggernaut were unlike anything I had dealt with before," he said. "The principal challenge—to give personality and meaning to a technology that was still being treated as though it were anonymous—interested me a lot. Also important was the fact that Apple offers a supportive environment. It's the kind of place where a designer can focus less on day-to-day business and more on design as a craft."[48]

But California was a long way from London. Trying to tip the balance, Brunner flew Jony and his wife, Heather, out to California for a second visit. On returning to London, however, Jony still couldn't make up his mind.

The opportunity wasn't a secret. Grinyer, Darbyshire and Phillips pushed him to move to California. 'We all said, 'It's a great opportunity Jon, how can you not take it?'" said Darbyshire.[49] Phillips pointed out that the other partners all had children. They were "tied to London, and he [wasn't], so it was a no brainer."[50]

Ultimately, none of the Project Juggernaut designs would see the light of day. Even so, the project was instrumental in moving Apple away from its previous beige boxes. Key ideas like split keyboards and docking stations (an idea that later came to market as the duo dock), as well as the gray and black industrial design that Apple became famous for in the early 1990s, were by-products of Juggernaut thinking.

It also became increasingly clear that, at least in part, Project Juggernaut was an elaborate ruse by Brunner to recruit Jony. "We always suspected that the project was given to Tangerine to try and lure Jony to the sunny climes of California to see if they could poach him over to the States," said Phillips.

Finally, Jony made the call. As he remembered, "through some sort of reckless sense of faith," he got to Yes.[51]

Brunner had tried to hire Jony three times: when Jony first visited Lunar as a student, when Brunner first started at Apple and when Jony had just joined Tangerine and after Project Juggernaut. "He liked California," said Brunner. "He liked the energy. So we managed to hire him on the third try. And that's how you do it. You identify the people you think are great and work on it until you eventually get them."

Another factor was undoubtedly that Jony was frustrated with consulting. He had achieved what many designers dream of: a successful practice with a lot of freedom. But consulting also restricted his ability to truly make an effect. "Working outside a company made it difficult to have a profound impact on product plans and to truly innovate," he said.[52] In most cases, by the time his commissions had been accepted, many of the critical decisions had already been made internally. Jony had come to believe that to do something fundamentally new required dramatic change from within an organization.

"While I had never thought that I could work successfully as part of a corporation—I always assumed that I would work independently—at the end of this big program of work for Apple, I decided to accept a full-time position there and to move to California."[53]

Tangerine would continue to thrive without him, and work on

projects for Apple, Ford and LG. The company is still running with Darbyshire in charge. In recent years, they are best known for their work on the British Airways flat bed, an innovative first-class seat that converts into a bed. Still, Jony Ive has gone on, left them for the world of Apple. His former partners remained philosophical, understanding that Jony couldn't be kept back. Brunner tried to ease the loss too. Phillips reported, "Bob was very kind when Jon had gone and gave us a nice big beefy project to boot as a bit of a sweetener. It was a nice chunk of money to say 'sorry.'" [54]

CHAPTER 4

Early Days at Apple

> I can't have people working in cubicle hell. They
> won't do it. I have to have an open studio with
> high ceilings and cool shit going on. That's just
> really important. It's important for the quality of
> the work. It's important for getting people to do
> it.
> —ROBERT BRUNNER

In September 1992, at age twenty-seven, Jony accepted a full-time position at Apple. He flew to California with his wife, Heather. The couple moved into a modest house on San Francisco's Twin Peaks, the highest hill in the city, from which they enjoyed a stunning view of the city that extended the length of Market Street to the skyscrapers downtown.

Inside, the place reflected Jony's design tastes. "There is a fireplace in the sparsely appointed interior and a tiny television sitting atop an upscale stereo with a turntable, and virtually all the furniture is on wheels," wrote reporter John Markoff, who visited Jony and his wife for a *New York Times* profile a few years later.[1] "The room is lighted by a futuristic lamp, which appears to hang like a red orb, but there isn't a personal computer in sight."

Jony bought an orange Saab convertible for the commute to Apple, about thirty-five miles away down the Peninsula in Cupertino. He went to work in the IDg studio on Valley Green Drive, a short walk from Apple's main campus on Infinite Loop. The IDg studio was Apple's first and Robert Brunner's brainchild. The choices Brunner made in setting it up were auspicious and they'd have a wide-ranging effect at Apple (that would be especially true after Steve Jobs's subsequent return).

Previously Apple had contracted most of its design to Frog Design, a full-service design consultancy run by Hartmut Esslinger, a hotshot German designer. Esslinger had developed a unified design language for Apple called "Snow White" that catapulted the company to the top of the ID world. However, by the late 1980s, Frog was starting to get expensive. Its Apple billings rose to more than two million dollars a year, twice what it would cost to use most other outside design firms, and much more expensive than running a small in-house design team. But Apple was stuck with a contract that Steve Jobs had negotiated with Esslinger in the early eighties, and couldn't get out without paying a huge penalty.

But money wasn't really the problem. By that point, Apple had more money than it knew what to do with. The company was making vast sums riding the desktop publishing revolution. Thanks to the Mac's graphical user interface, great layout software and cheap laser printers, Apple was selling boatloads of machines to newspapers, magazines and book publishers. At the end of 1988, the company had three factories working twenty-four hours a day, seven days a week. Yet, despite a research and development budget of two hundred million dollars, Apple had no disciplined product pipeline.

Different product groups—peripherals, portables, desktops and so on—did have the next generation of products in the works, but nothing beyond. Without any coordination between groups, Apple was reverting to the pre–Snow White days when each division had its own design ideas. Products from the printer group looked nothing like those from the monitor division. It was as though Apple was four or five separate companies. The company needed a regimented product pipeline and a new, unified design language to give the products cohesion.

By 1987, there was a general recognition within Apple that the answer was an internal design team, but without a visionary like Jobs, the engineers had no idea where to start; since design had been outsourced to Frog Design, there were no designers on staff. Apple's management first tried to find a superstar designer to replace Esslinger, thinking the firm needed someone with a world-class reputation.

In early 1988, they embarked on a global tour of the world's most famous design studios on a quest to find a spectacular talent. They traveled to Europe and Asia, interviewing design firms like Porsche Design Studio. They visited top designers in Tokyo, London and Berlin, but none were quite right. In Italy, they visited Mario Bellini (the so-called crown prince of Italian design), who rudely dismissed them. Not to waste the trip, a meeting was set up with Italian designer Giorgetto Giugiaro, who in the next decade would be named "Car Designer of the Century." Hired by Fiat at age seventeen, he was responsible for scores of gorgeous Bugattis, BMWs, Maseratis and Ferraris and was synonymous with Italian style, having designed more cars than anyone else in the twentieth century.

Apple executives visited Giugiaro in his giant, factory-like, ultra-secure Italdesign studio near Turin. They found him sketching with one hand, talking on the phone with the other, continually issuing orders to his many minions. The executives were so impressed, they gave "Il Maestro" a million-dollar contract to design concepts for just four products, which they wished to use as models for a whole line of products. Their hopes soon fell flat.

Accustomed to designing cars, Giugiaro worked from the outside in. He would make loose, impressionistic sketches of cars, which his model makers would use to make 1:1 clay models. Often, the finished models would differ quite a lot from the sketches. Over several months, the Apple engineers discovered that the Italdesign model makers made many design decisions; a Giugiaro sketch was more inspiration than blueprint. That was the opposite of the way things were done in California, so when Giugiaro applied the same methods to the Apple commission—his model makers designed the computer casings using clay, just as they would high-end Italian sports cars—the lack of regard to the internal components meant the models didn't translate to manufacturable products.

Just when the search seemed fruitless, it turned into borderline farce. The Apple team went to see hotshot Swiss-born German designer Luigi Colani, one of Jony's design heroes, who was famous for eccentric "biodynamic" designs of cars, motorcycles and consumer goods. After delivering a lecture at the Art Center College of Design in Pasadena, Colani was asked a question about the future of keyboards and he launched into a long diatribe in which he compared keyboards to women's bottoms. Since men like to grab women's rears, he said, the keyboard should be split up the middle to accommodate different hand sizes. To illustrate his anatomical theory, he drew a woman's bottom with keys on it and handed the sketch to one of Apple's staff, who was almost too embarrassed to take it. Back at Apple HQ, the story spread, and one of the staffer's colleagues brought in a female mannequin and stuck keys to it, replacing his regular keyboard. The stunt drew outrage from Apple's female staff, but, remarkably, the idea remained and Apple's computers soon had split ergonomic keyboards.[2]

While the upper management searched for a design genius, some of Apple's product groups had been working closely with Bob Brunner at Lunar. The projects included some off-line blue-sky product explorations like the Juggernaut project that Jony later worked on. Reportedly, Lunar's invoices classified the work as "product design," which indicated engineering rather than industrial design. In light of the exclusive contract with Frog, Apple's accounting department wouldn't pay for any industrial design work unless the invoice was in Frog's name and vendor number.

Meanwhile, as Apple's relationship with Esslinger soured further, less and less work went his way. The billings dropped precipitously and eventually Apple stopped paying Frog's retainer. At the same time, Esslinger was being pressured by Jobs to work at NeXT, which would violate Frog's contract with Apple. Ultimately, Apple and Frog agreed to nullify their contract.

That put Apple in a difficult spot. The company had dumped its design firm and its search for a new one had been a bust. Only then did someone at Apple realize they had their superstar designer right under their noses. Brunner had been doing stellar work, and the company had been delighted with everything he'd worked on. More than that, Brunner was more involved than the typical outside contractor. He regularly attended design meetings, even those that weren't directly involved in his own contract work, and he had been pushing Apple to mature Snow White and transition to a new, unified design language.

Apple started trying to recruit Brunner. They twice offered him the job of design director, but he wasn't interested because Apple had no design organization. He felt the job would be a dead end.

"I didn't want to work at a company that wasn't designing its own stuff," said Brunner. "I didn't want to manage people doing the creative work; I wanted to *do* the creative work."[3] By 1989, Apple was getting desperate and tried again, this time asking Brunner, What would it take?

Brunner was tempted to take the offer. "Of all the companies in the world, this company could have a really amazing in-house design team," he said. "It should have a really amazing studio. It has great products. A great brand. Great history."

He made a pitch. Brunner told Apple that he wanted to build a team within the company and turn Apple into a world-class design company.

But he didn't want to build a big, sprawling design organization, which was typical of companies Apple's size. He'd worked with some big brands and found that their design organizations were too big and multiarmed to do really creative work. Because they were big, they tended to be bureaucratic, another obstacle to good design.

Instead, Brunner wanted to re-create his small design agency, Lunar, within Apple. He wanted a "small, really tight" studio. "We would run it like a small consulting studio, but inside the company," he said. "Small, effective, nimble, highly talented, great culture."[4]

Setting up a consultancy inside Apple seemed in line with the company's spirit: unconventional, idea driven, entrepreneurial. "It was because, really, I didn't know any other way," Brunner explained. "It wasn't a flash of brilliance: that was the only thing I knew how to do."

Apple agreed. Brunner, then thirty-two, joined the company as head of ID in January 1990. But the job wasn't anything like he'd imagined. He was head of the design studio—and its only member. He was given a desk in the middle of the hardware department.

"I got there and it was, here's your cubicle, in a sea of engineers. I thought, 'Oh God, what have I done?'"

Brunner's Dream Team

Despite his intentions to build a dream team, about eighteen months passed before Brunner began hiring in earnest. He needed to make the case for more resources with Apple executives but, perhaps more important, he had to come up with a cool place for the designers to work.

"[The studio] was essential to recruiting talent," Brunner said. "I can't have people working in cubicle hell. They won't do it. I have to have an open studio with high ceilings and cool shit going on. That's just really important. It's important for the quality of the work. It's important for getting people to do it."[5]

Brunner found part of the answer in an underutilized building that Apple was leasing at 20730 Valley Green Drive. Called Valley Green II, or VGII, the building was a large, low-slung Spanish-style stucco structure surrounded by a few small trees and a big parking lot. Not far from Apple's main campus, Valley Green Drive is on the other side of De Anza Boulevard, the main road through the center of Cupertino. Almost all of the buildings in the area are leased by Apple, making this part of Cupertino look like a company town. Apple's first office, on Bandley Drive, is just around the corner.

Brunner took over half of the building, a big open space with twenty-five-foot-plus ceilings. He would share the building with Apple's Creative Services group, known as Apple's "In-House Design Consultants," who were called on to produce things like brochures, manuals, in-store posters and displays and video promos.

Another crucial consideration for Brunner was that the building was not directly under the noses of Apple's meddlesome executives. "I liked that it was off the beaten path."

Brunner worked with Studios, a big San Francisco architecture and design firm, to transform the interior into an appealing design studio. Apple's practice was to use standardized office furniture from Herman Miller, the company credited with inventing the cubicle, but Brunner didn't assemble it into cubes. Instead, the desks were arranged in unusual formations around the space. "We used the taller structures as spines running around the studio and the work spaces running off it," said Brunner. "The corporate planning people didn't get it. They said you can't do that, but that's what we did, put it together in different ways. It totally freaked them out. It was great, a lot of fun. We just made it less oppressive."

Brunner had a CAD workstation installed for creating 3-D models of designs along with a computer numeric controls (CNC) milling machine to turn the CAD models into foam ones and a paint shop for testing different colors.

"The ID studio at Apple was a cool work space," recalled Rick English, a photographer who did a lot of work with Apple in the 1980s and 1990s. In 1997, English contributed photos to Kunkel's book about the design group, *AppleDesign*, but he also worked with a lot of other design studios in the Valley. To his eye, Apple seemed different. It wasn't just the tools and their focus; the place was rapidly populated with designer toys, too, including spendy bikes, skateboards, diving equipment, a movie projector and hundreds of films. "It fostered this really creative, take-a-risk atmosphere, which I didn't see at other firms," said English.[6]

When Brunner began recruiting in earnest, he initially found attracting talent difficult. Apple had no reputation for doing its own design, having outsourced it to Frog, and talented, ambitious designers were more inclined to go to firms with a strong creative history like the Bay Area's IDEO.

To help with recruiting, Brunner took a leaf from Tangerine's playbook and started promoting his work through design magazines. He created mock-ups of fantastical Apple products and ran big glossy photos of them on the back of *I.D.* magazine, the international design bible. One was a gigantic bicycle navigation computer that showed maps and local landmarks on a black-and-white screen. Another was a chunky wristwatch computer the size of a cantaloupe.

"They were concepts, not real products," said Brunner. "They started to get attention. It was totally recruiting. No other reason. They were sketchy, information appliance models. A little bit tongue in cheek, but it served its purpose."[7]

Over time, Brunner recruited a team of talented designers, some of whom would remain with Apple for decades and be responsible for a string of hit products, including the iPhone and iPad. Key members of Brunner's team were Tim Parsey, Daniele De Iuliis, Lawrence Lam, Jay Meschter, Larry Barbera, Calvin Seid and Bart Andre.

Daniele De Iuliis (day-YOU-lease) was perhaps the most precocious of the group. Born in Bristol, United Kingdom, of Italian descent, De Iuliis was a graduate of the Central Saint Martins College of Arts and Design in London. Brunner hired him in 1991 from the San Francisco office of design group ID Two (where Jony's friend Clive Grinyer had earlier worked).

Brunner specifically wanted designers with experience in consulting. "After experiencing the inertia that exists inside Apple, Bob realized that by hiring former consultants he could operate with the speed and efficiency of a freelance group," said Tim Parsey. "As a former consultant, Bob knew that we would think and act like consultants."[8]

Fellow designer Barbera was impressed with De Iuliis's personality right away: "Danny, in particular, gave off that weird light that other designers tend to notice. I took one look at him and figured that our work was gonna get a lot better, fast."[9]

De Iuliis was able to imbue his designs with strong personality, a skill that served him well later on. One of his early projects was the Macintosh Color Classic, an update of the original Mac that exuded character and was avidly collected by fans for years. He would later work on the MacBook Pro and the iPhones 4 and 5. His name appears on more than 560 patents. They're vast and varied in scope, including innovations in 3-D cameras, multi-touch displays, location tracking, RFID transponders, nitriding stainless steel, magsafe charging mechanisms, the iPod and improved speaker enclosures.

Later in his career, De Iuliis would receive top design awards for his work. Once Jony joined the team, the two developed a strong relationship. De Iuliis and Jony lived close to each other in San Francisco, and commuted together for more than twenty years.

In 1992, Brunner recruited Bartley K. Andre (known as Bart), a graduate of the University of California at Long Beach and an intern in Apple's Personal Intelligent Electronics, or PIE group. He would emerge as one of the top five patent holders in the United States on a year-to-year basis (thanks to his last name, he is listed on all of Apple's major patents in the title: "United States Patent Application Andre et al."). By 2013, Andre had more patents to his name than any other Apple designer, including Jony. In 2009 alone, he received 92 patents; in 2010, his 114 set a record for an Apple designer. Most of the patent awards were for innovations on the phone, tablet and laptop lines.

Andre worked on everything at ID, from circuit modules to RFID systems. He was credited with the design of Apple's 035 design prototype of the first iPad, according to information released during the *Apple v. Samsung* trial in 2012. Along with other members of the team, he several times received the prestigious Red Dot Award, from Germany's Design Zentrum Nordrhein Westfalen institute.

Daniel J. Coster joined the team after Jony, arriving in June 1994. Described as "tall, goofy [and] super-talented," Coster had earned an ID degree from the Wellington Polytechnic School in New Zealand in 1986. Initially hired on a three-month contract, he impressed the group with work on colors and finishes for the Newton portable, then was hired full time. He designed various towers and gained notice for being the lead designer of the Bondi Blue iMac. Like his coworkers, Coster rapidly accumulated patents, receiving nearly six hundred working for Apple over the last two decades. In 2012, Coster was inducted into his alma mater's design Hall of Fame for "an outstanding contribution to New Zealand's economy, reputation and national identity through art and design."[10]

When Jony joined the team, his arrival was much anticipated. "Bob knew the effect that a strong new designer would have on the group," said Meschter. "When Daniele De Iuliis and Tim Parsey first arrived, our whole approach to design changed. But when Jonathan came on board . . . the group really took off."[11]

To run the group like an outside consultancy but within Apple, Brunner set up a loose management structure that largely persists today. The designers always worked together on whatever project the group was working on. "We'd work on multiple projects, and move from project to project, pretty much the way he [Jony] does it today," Brunner explained.[12]

Brunner also made about half a dozen of the designers "product line leaders" (PLLs) for Apple's major product groups: CPUs, printers, monitors and so on. The PLLs acted as liaisons between the design group and the company, much in the way an outside design consultancy would operate. "The product groups felt there was a contact within the design group," Brunner said. "They managed the communication and the different needs of each group. I didn't know any better. I ran it like the Lunar studio. There'd be discussion of a project. We'd design it, make it and ship it."[13]

As Apple grew and thrived, riding the desktop publishing revolution and the exploding market for PCs, the group found itself with a massive workload. "We had an enormous number of products in the pipeline," said Brunner. "Two lines of desktops, monitors, printers, mobile products. Way too many. It was an enormous amount of work. More than we could handle."

The production schedules also got shorter and shorter. When Brunner first started at Apple, the product development cycle was eighteen months or more. "It was crazy generous," Brunner said. "You had an amazing amount of time to make something work." Within a couple of years, however, the product development cycle shrank to twelve months, then nine, and sometimes even six months if the product was needed in a hurry.

"All of a sudden, what got compressed was our thinking time," Brunner said. "It still took just as long to implement something, but the time to explore, to test and to play with, just went away."

Another challenge to Brunner and his crew was that Apple's internal culture heavily favored the engineers within the product groups: The design process was engineering driven. In the early days of Frog Design, the engineers had bent over backward to help implement the design team's ambitions, but now the power had shifted. The different engineering groups gave their products in development to Brunner's group, who were expected to merely "skin" them.

Brunner wanted to shift the power from engineering to design. He started thinking strategically. His off-line "parallel design investigations" were a key part of his strategy. "We began to do more longer-term thinking, longer-term studies around things like design language, how future technologies are implemented, what does mobility mean?" The idea was to get ahead of the engineering groups and start to make Apple more of a design-driven company, rather than a marketing or engineering one. "We wanted to get ahead of them, so we'd have more ammunition to bring to the process."[14]

For every skin job Brunner did for engineering, he'd launch up to ten different parallel design directions. It sometimes seemed that such investigations were set up as competitions. He'd have several designers—inside the group and out—submit initial concepts. "It was almost like a competition, which Bob encouraged," said English. "Then, when one of those designs was selected it became that designer's baby all the way 'til completion."

The Juggernaut project with Jony at Tangerine had been one such parallel investigation, and several others unfolded with outside agencies like Lunar and IDEO. (This practice continues to this day at Apple, although both Jony and Steve Jobs avoided admitting it publicly.) The parallel design investigations also allowed the overworked design team to work with talented designers not on the Apple payroll. "Sometimes we wanted to hire specific designers to act as part of the team, to use them as freelance talent," Brunner said.

Brunner was good at getting attention for the group, which inevitably led to design awards. Every month, Brunner ran ads on the rear cover of *I.D.* magazine. No longer limited to blue-sky concepts, Brunner published photos of designer prototypes, mostly to give his designers attention and make them feel good. It was an expensive motivational tool (Rick English said he was billing the company at least $250,000 a year in those days), but some of the work that the team had done was soon displayed in the studio as big photos on the wall.

Everything was documented. English and another photographer, Beverley Harper, photographed all the finished designs and a lot of the concepts. As they moved away from beige boxes, the designers felt the work deserved to be recorded. "The mind-set was that there was going to be a historical archive of all the things they worked on," said English. "They absolutely believed that their work had such importance." The habit lived on, as Jony's design group continues to document everything it does today.

In hindsight, Brunner's choices—the studio's separation from the engineering groups, its loose structure, the collaborative workflow and consultancy mind-set—turned out to be fortuitous. One of the reasons Apple's design team has remained so effective is that it retains Brunner's original structure. It's a small, tight, cohesive group of extremely talented designers who all work on design challenges together. Just like the designers had done at Lunar, Tangerine and other small agencies. The model worked.

Jony to the Rescue

Jony's first big assignment at Apple was to design the second-generation Newton MessagePad. The first Newton hadn't yet been released, but the design team already hated it. Thanks to a rushed production schedule, the first model had some serious flaws that Apple's executives, as well as the designers, were eager to fix.

Just before the Newton was shipped, Apple discovered that the planned lid to protect its delicate glass screen wouldn't clear expansion cards, which were to be inserted into the slot at the top. The design group was charged with developing some quickie carrying cases, including a simple leather slipcase, and off it went into the marketplace. In addition, the Newton's loudspeaker was in the wrong place. It was in the palm rest, so the user tended to cover it up when holding the device.

The hardware engineers wanted the second-generation Newton (code-named Lindy) to have a slightly larger screen for better handwriting recognition. Since the pen was attached awkwardly to the side, a kludge that gave the Newton extra width, they wanted the new version to be significantly thinner; the original was so bricklike, only the largest of jacket pockets could accommodate it.

Jony worked on the Lindy project between November 1992 and January 1993. To get a grip on the project, he began with its design "story"—that is, by asking himself, *What's the story of this product?* The Newton was so new and versatile and unlike other products, that articulating what it was primarily used for wasn't easy. It morphed into a different device depending on what software it was running, so it could be a notepad or a fax machine. CEO Sculley called it a PDA but, for Jony, that definition was just too slippery.

"The problem with the first Newton was that it didn't relate to people's everyday lives," Jony said. "It didn't offer a metaphor that users could grasp." He set about fixing that.[15]

To most people a lid is just a lid, but Jony gave it special attention. "It's the first thing you see and the first thing you interact with," Jony said. "Before you can turn the product on, you must first open the lid. I wanted that moment to be special."[16]

To enhance that moment, Jony designed a clever, spring-loaded latch mechanism; when you pressed the lid, it popped open. The mechanism depended on a tiny copper spring carefully calibrated to give just the right amount of pop.

To allow the lid to clear any expansion cards in the slot on top, Jony created a double hinge to allow the lid to clear any obstructions. When the lid was open, it flipped up and over the back to be stored out of the way. That conveyed something to the user too. "Pushing the lid up and around the back was important because the action is not culturally specific," Jony noted at the time.

"Folding the lid to the side, like a book, created problems because people in Europe and the U.S. would want to open it on the left whereas people in Japan would want to open it on the right. To accommodate everyone, I decided the lid would have to open straight up."[17]

Next, Jony turned his attention to the "fiddle factor," the special nuances that would make the product personal and special. The Newton was pen based, so Jony focused on the pen, which he knew users would love to play with. Jony's solution to the challenge of reducing width and integrating the pen into the MessagePad itself was a storage slot at the top. "I insisted the lid fold up and over the top, like a stenographer's notepad, which everyone understands [and] . . . users saw Lindy as a notepad. The stored pen at the top, where a stenographer's notepad's spiral binding would be, made the right connection.

"That became a key element of the product's story."[18]

The slot was too short for a full-size stylus, so Jony created a stylus that cleverly telescoped. Like the lid, the pen featured a pop-up mechanism that made it pop out when the user pressed its top. To give it weight and heft, he fashioned the pen from brass.

His colleagues all went nuts for it. "Lindy was Jonathan's shining moment," said fellow designer Parsey.[19]

On top of all this, Jony was under an extremely tight deadline with enormous pressures to deliver. The first edition of Apple's pioneering handheld device had been doomed by the *Doonesbury* cartoon that came to define it. Cartoonist Gary Trudeau depicted the Newton's handwriting recognition as hopeless, delivering a gut punch to the device from which it never recovered. Thanks to Trudeau, the first Newton MessagePad had to be replaced as quickly as possible.

The pressure fell to Jony. "When you're aware of the lost revenue each day the schedule slips, it tends to focus your attention," he said with typical British understatement.[20]

To the amazement of his colleagues, Jony was able to go from the initial design to the first foam concept models in two weeks, the fastest anyone had seen. Jony was determined to finish the project on time, and he traveled to Taiwan to fix manufacturing problems. He camped out in a hotel near the factory where the Newton would be made. He and a hardware engineer troubleshot the pen's pop-up mechanism in his hotel room.

Parsey remembered Jony pushing himself to create something special. "To do the best design you have to live and breathe the product. At the level that Jonathan was working, it becomes like a love affair. The process is exhilarating . . . and exhausting. But unless you're willing to give everything to the work, the design will not be great."[21]

When it was done, Jony's colleagues were stunned and impressed with both the new Newton and Jony, who had joined the team only months earlier. Apple executive Gaston Bastiaens, who was in charge of Newton, told Jony he would win every single design award. He nearly did. After Lindy's introduction in 1994, Jony won several of the top awards in the industry: the Gold Industrial Design Excellence Award, the Industrie Forum Design Award, Germany's Design Innovation Award, a Best of Category award from the *I.D.* Design Review and the honor of being featured in the permanent collection of the San Francisco Museum of Modern Art.

One of the things about Jony that struck Rick English was Jony's dislike of awards. Or, rather, his dislike for receiving awards in public. "Even early on, Jony Ive stated that he was not going to go to those events," said English. "That was interesting behavior because it was really different. He hated going up on stage and receiving awards."

Jony's Newton MessagePad 110 was on the market by March 1994, only six months after the original Newton went on sale. Unfortunately, no amount of fiddle factor was enough to save the Newton, as Apple made a series of blunders marketing it, both rushing the first device to market before it was ready and hyping its capabilities. In the face of unrealistic expectations, the Newton never reached critical mass. Both generations of Newtons were also plagued with battery problems and the poor handwriting recognition that Trudeau mocked. Not even Jony's stellar design work could save it.

Phil Gray, Jony's old boss at RWG, remembers seeing Jony in London just after his MessagePad 110 came out. "The Newton was like a brick in retrospect, but at the time was a handheld device that no one had done before," Gray said. "Jony was frustrated because although he had worked really hard on it, he had to make a lot of compromises because of the engineering elements. Afterwards, at Apple, he went on to be in a position where he not only could influence engineering but also manage and control those processes."

The MessagePad also marked an important transition in Apple's manufacturing strategy. The MessagePad 110 was the first Apple product outsourced entirely to Taiwan. Apple had partnered with Japanese companies before (Sony for monitors, Canon for printers), but generally made its products in its own factories. For the MessagePad 110, Apple outsourced the Newton to Inventec. "They did a really amazing job, it went really well," Brunner said. "The quality turned out to be really high. I credit Jony with that. He basically broke his back, spent an enormous amount of time in Taiwan getting that thing just right. It was beautiful. Well executed. It worked really well. It was an amazing product."

That decision initiated a growing reliance on outside contractors to build Apple products, a practice that would become controversial a decade later.

Soon after completing the Lindy project, Jony had an idea to simplify the design of Apple's bulky cathode-ray tube (CRT) monitors, which were perhaps the least sexy of Apple's products and among the most expensive to manufacture. Because of their size and complexity, the molds for each of the plastic monitor housings—and there were dozens of models at the time—could cost more than a million dollars to tool.

To save money, Jony came up with the idea of a new case design with interchangeable parts, which could be adapted for several monitor sizes. Previously, monitor cases came in two parts: the bezel (a face frame that cradles the front of the cathode-ray tube) and the bucketlike housing that encloses and protects the CRT's back. Jony's idea was to split the case into four parts: the bezel, a mid-bucket and a two-part back bucket. The modular design would allow the mid- and back bucket to remain the same across the product line. Only the front bezel would come in different sizes to accommodate different-sized monitors.

In addition to saving money, the new case would be better looking, its trimmer design allowing for a tighter fit around the different CRTs, making them appear smaller and more sculptural. Jony's design introduced a couple of new elements into the group's design language, including new treatments for vents and screws. "The new approach is more subtle,"[22] said designer Bart Andre, who designed the actual enclosures based on Jony's idea. His work seemed to attract everybody's attention.

Off to a Running Start

Although he wasn't hired as a manager, Jony stood out as a natural leader. "Jony Ive was very serious about his work," English remembered of those early days. "He had a ferocious intensity about it. He was calm, but very deep. He was very serious, but also a really nice guy. He led in a quiet way: he inspired people to work for him."[23]

Jony began to emerge as Brunner's second in command. Not only did he provide ideas and design taste, he soon helped recruit the next group of designers. Within a couple of years, Jony hired most of the rest of the team that would go on to make the iMac, the iPod and the iPhone, including Christopher Stringer, Richard Howarth, Duncan Robert Kerr and Doug Satzger.

Christopher Stringer, born in Australia in 1965, had been raised in the North of England. He attended North Staffordshire Polytechnic in Stoke-on-Trent and graduated from London's Royal College of Art in 1986. A veteran of IDEO—hired in 1992, he helped develop Dell's design language and won an ID Design Review Award for an innovative light switch—he was recruited by Jony in 1995 as a senior industrial designer.

Stringer worked on the early PowerBooks and tower computers. Over the next seventeen years, he would be involved in all the major releases (including the iPhone), peripherals and in even smaller projects, like the design of product packaging. He was also the first designer to give testimony at the *Apple v. Samsung* trial, where, according to Reuters news service, "Stringer looked every inch the designer with his shoulder-length hair, salt-and-pepper beard, wearing an off-white suit with a narrow black tie."[24] Stringer was often seen at Apple launch events talking side by side with Jony. The impression that they are close friends is enhanced by their shared history; both hail from Staffordshire and studied in the north of England.

Richard Paul Howarth was born in Lukasa, Zambia, and graduated from Ravensbourne College of Design and Communication in London in 1993. He was recruited by Apple in 1996 from IDEO and became one of the group's main designers. Howarth was the lead designer of the original iPhone, and a major contributor to the iPod touch and iPad.

Another designer from the United Kingdom, Duncan Kerr graduated from the Imperial College London in 1985 with a degree in mechanical engineering and a degree in ID engineering from the Royal College of Art. He was also recruited from IDEO. As one of the team's more technical members, Kerr has great influence in the development and investigations of new products and technologies. He helped pioneer the multi-touch technology that led to the iPhone and iPad. He has been named in numerous patents, including various technical innovations involving components like proximity detectors, display modules and magnetic connectors.

Doug Satzger was the fourth IDEO alumnus. Satzger attended the University of Cincinnati and graduated in 1985. He started his career at IDEO as an industrial design lead, before moving to design TVs at Thomson Consumer Electronics. He would work at Apple between 1996 and 2008 in the IDg. An Ohio native, his interest in materials and knowledge of manufacturing processes made him the group's design lead for color, materials and finishes, working on the first iMac to the latest iPhone, iPods, iPad and MacBooks. Satzger has been named in many patents, mostly in electronic devices, displays, cursor controls, packaging and connectors. (After Apple, Satzger joined HP/Palm as the senior director of ID then moved to Intel, where he's vice president of the Mobile and Communications Group and general manager of ID.)

When the design team looked to recruit a new designer, engineering and computer skills were a plus, but not absolutely necessary. "We are looking for personality, overwhelming talent and the ability to work in a small group," said De Iuliis. "We also want to be impressed with a designer to the point of intimidation."[25] To put it another way, the group was more likely to hire a talented car designer than a mediocre computer guy.

Another key member of the group in the mid-1990s and after was Calvin Seid, a native of Portland, Oregon, who graduated from San Jose State University in 1983 and worked for design firms in Oregon and Silicon Valley after graduation. He joined the Apple IDg in 1993 to design and manage CPU projects. (He died unexpectedly on April 6, 2007, of coronary artery disease at the age of forty-six. He was popular, and his death upset the team greatly.)

Though distinctly international, the design team profile was largely white and male. With the exception of Seid, who was of Filipino descent, the members of the team were all young white guys, most of them from the United Kingdom. There was one woman in the 1990s; by 2012, there were two women on the team of about sixteen designers.

"Jony's had that core of people around since then. They've been commuting up and down Highway 280 together for twenty years now, between San Francisco and Cupertino," said Sally Grisedale, who worked closely with Jony's group in the late 1990s. "They're tight. They're family. Many started as single men and then they had families and now all live in the same neighborhood."

The IDg was a great place to work; it seemed like nobody ever quit. But the lack of turnover was actually a challenge. Jony would admit to complicated feelings about the stability of the team. "Though we don't want people to leave the group, the lack of movement makes it difficult to bring in fresh talent," he said. "We need new people at regular intervals to prevent ourselves from stagnating. But this can only happen if other people are willing to leave."[26]

The Espresso Aesthetic

With new and gifted designers in place, the Apple IDg started working on a new design language for the company's products. The aging Snow White language no longer suited the growing range of Apple products. The off-white or gray color schemes, with lots of horizontals on the enclosures, seemed ill adapted to the plethora of new printers, handhelds, speakers and portable CD players.

The team came up with what it called "Espresso," a Euro-style aesthetic characterized by swooping organic shapes, bulges and an adventurous use of colored and textured plastics. Less a design language per se than a loose set of guidelines and best practices, Espresso was, in short, an aesthetic. There were no hard-and-fast rules. But, as has been said of pornography, the designers knew it when they saw it.

The Espresso name has two possible origins. The official story is that it was inspired by the minimalist design of the modern European coffee pots the group used while working. The unofficial (and more likely) story comes from Don Norman, head of Apple's Advanced Technology Group in the mid-1990s.

"The name was a derogatory term applied to the new design team, who had just installed a fancy espresso coffee machine in the studio. One old-time engineer said it was a sign of the 'yuppification' of Apple, and started calling the team 'espresso.' The funny thing is, the designers didn't get it and adopted the term for their new design language."

One of the first Espresso products was the Macintosh Color Classic, an update of Steve Jobs's original Mac, for which De Iuliis is given credit. Like the earlier Mac, it was an all-in-one machine, but De Iuliis lengthened its face, made the vents look like gills, gave it a higher forehead and made the floppy slot even more mouthlike. More bulbous and curvy than the original, the Mac Color Classic came out with a distinct personality. Enthusiastic users went crazy for it and turned the machine into a highly collectable machine.

Its most distinctive Espresso touch was a pair of small, round legs at the front, which looked like the feet of a baby elephant. The fat feet tilted the computer upward by six degrees. As Don Norman put it, they gave the machine the look of "an eager pet staring up adoringly at its owner." In fact, the feet were a kludge, the result of a lucky break. "Came from a fuck up," explained Norman.[27]

"Early pizza box machine was about to go into production [and we] focused so much on making it slim and flat we forgot about the floppy slot on the front. Not enough room to insert a disk with a keyboard in front of the machine. So we added a pair of feet at the front that tilted the box up. Had the unexpected effect of giving the machine a lot of personality . . . [and it] became a design feature that was featured prominently for five years." The look would influence a later generation of smash-hit products even after Jobs returned, as the iMac has its origins in Espresso.

Project Pomona

Jony's next big project was the Twentieth Anniversary Macintosh, which would be the first major project to get started within the design group rather than one of Apple's engineering groups. "At its best, engineering and design would work hand in hand," Brunner explained. "Other times they would come to us and say, This is the product, just make it look pretty. It was already defined, you just need to put your styling on it. That was Apple at its worst."

Brunner wanted Project Pomona to signal a shift in development.

"This was one when it wasn't engineering driven at all. It was design driven. It was all about a certain type of experience that we saw and thought was important."

Launched in 1992, Project Pomona would be one of Brunner's parallel design investigations. Just like the Juggernaut project, Pomona involved the whole IDg, along with a few freelance designers. The ambition was large: Project Pomona aimed to imagine the first computer designed for the home, rather than the workplace. The end result would be a triumph—and a disaster.

By the early nineties, more and more computers were being used in people's homes, but they were mostly beige boxes that had been designed for office cubicles. Brunner wanted to change that. "For years I wondered how the computer would evolve from a box into something more physically compelling that would fit better in the home," said Brunner. His hope was that his team would come up with "concepts that would encourage people to select their computer the same way they would a piece of furniture or a home stereo."[28]

Brunner also wanted to move away from the heavy, oversized CRT monitors standard on desktop computers. Instead, he wanted to fuse a desktop CPU with a flat-panel display. "We thought that flat panels would become mainstream; they were already mainstream on laptops."

Brunner's October 1992 briefing document laid out his ideas and criteria for a high-design desktop Mac. It was, in effect, a challenge to the group's designers and five outside consultants to come up with the best concepts.

Brunner kept it loose: His basic call was for a high-design desktop Mac, powerful but with a minimum footprint. Brunner insisted that all concepts use new materials in new ways, including polished or brushed metal, wood, veneer and different types of coatings and finishes. Not only were there a minimum of other restrictions; the designers were actually invited to step outside of Apple's established design language.

Brunner did add another interesting wrinkle to the project: He wanted a machine that couldn't be expanded with extra hardware cards and beefier internal components. Most home users never bothered to expand their machines, so he encouraged designers to forget expansion slots, freeing them to explore much thinner designs.

The initial concepts were wildly varied. One was inspired by the design of a classic Tizio lamp, with the guts housed in the base and the screen mounted on an arm that hovered in space. Another concept hid the main display and components inside a metallic exoskeleton.

One of the most intriguing concepts came from Jony and Daniele De Iuliis, who teamed up to pitch a mid-range computer. Their design had a homely look and their goal was to create a machine affordable for those with a modest budget. They called it the "Domesticated Mac."

To keep the price down, they based it on a CRT monitor, not a pricier flat screen. It was basically a Classic Mac in a funky-looking case. It was an odd duck, resembling an old-fashioned wardrobe, with three feet and twin doors that covered the display. There were slots inside the doors for things like extra floppy disks. Jony and De Iuliis also put an analog clock in one of the doors. Cleverly installed, the clock would flip around so that it told the time when the door was open or closed.

Brunner created his own Pomona project design. His concept closely aligned with his prescription for a futuristic computer with a slim profile and powerful components. Brunner designed a wide, curved enclosure containing a flat-panel display flanked by a pair of big stereo speakers. It was a computer-cum-stereo, perfect for the kind of multimedia experience promised by CD-ROMs, then new to the market. To keep it slim, he proposed to use the guts from a PowerBook notebook. It would be made from—of all things—black mahogany, like a concert piano.

Since the other designers thought his concept looked more like a product from the high-end audio maker Bang & Olufsen than a PC, Brunner's solution became the "B&O Mac." The mating of a PC and stereo system was a novel idea at the time, and it generated a lot of excitement in the design studio. In fact, Brunner's concept would trounce all other Pomona designs in focus groups in the summer of 1993 and, by the end of the project, was declared the winner of the Pomona competition.

Nearly a year had passed since Brunner released his brief, but the group had a good idea of the basic shape and scope of the project. So far so good.

To turn it into a real product, Brunner handed the project over to Jony in the summer of 1993. Jony had just finished his work on the Lindy MessagePad 110 and, when handed the B&O Mac by Brunner, he knew he was facing a tough challenge. Going back to basics, he started with the design story.

"On a technical level, we understood the challenges associated with packaging a lot of components into a very slim space," Jony recalled later. "But philosophically, the project was more challenging. Like the first Macintosh, the design had no predecessors, which meant I had to come up with a new meaning for the product. I wanted the design to be simple almost to the point of being invisible."[29]

Ultimately, Jony would keep the spirit of Brunner's concept but change almost everything else. He redesigned the proportions of the computer. Where Brunner's initial design was wide and curved and appeared to take over a desk, Jony made it taller and much narrower. He changed the size of the foot of the base (which was called the bale) and created a hinge that allowed the foot to double as a carrying handle. Handles would feature prominently in Jony's designs. He redesigned the back panel, giving more room to the CPU and motherboard.

In April 1994, after working on it all winter, Jony handed over his design to a pair of product design engineers to make a working prototype. As the prototype took shape, a marketing manager worked up an internal product brief. The machine gained the official code name "Spartacus." After eighteen months, everything was on track to turn it into a real product.

Then Spartacus stumbled into its first major hurdle. To keep it slim, Jony had planned to use components from the portable PowerBook, only to find the working prototype was seriously underpowered. Portable components then lagged behind their desktop counterparts by at least a generation, so Spartacus seemed painfully slow. In particular, video was poor and, because of the flat profile, the circuitry couldn't be enhanced with a souped-up video expansion card like its cousins in the desktop department. This loomed as a major liability because Spartacus was to be sold as a desktop computer. Consumers would expect nothing less than desktop performance.

Jony moved to a regular desktop circuit board (developed for the Performa 6400), but a new problem arose when the marketing department told him that no one would buy a desktop machine unless it could accommodate expansion cards. Even if Brunner was right about home users, the marketing experts advised that an un-expandable desktop would be commercial suicide. To accommodate a pair of expansion cards, Jony was forced to design a special clip-on "hunchback" that covered any cards the user might add. Dubbed internally the "backpack," the add-on would ship with every machine.

"With the original back in place, the design is powerful yet physically lean," Jony would say. "But with the backpack inserted, it becomes a real power system, expressing on the outside the enhanced function contained on the expansion card inside."[30] That may sound like designer doublespeak, but Jony tried to put a brave face on what was obviously a horrible kludge.

Despite the misshapen hunchback, the design team became quite excited by the prospects of Spartacus. For an executive presentation in 1994, the group rustled up bigger and smaller versions, showing how the concept might be extended into a whole family of desktop products.

At every step, they faced resistance from the engineers. "There were layers and layers of middle managers, many who had come from Dell or HP and didn't understand the design-driven approach," Brunner explained. "They were accustomed to slapping a cheap metal skin on a product, because that was the way they did it at Dell, and Dell sold a lot of computers. They didn't really believe in what we were doing, and the very senior management of the company at the time didn't step in. So that made for a fight at the second and third levels."

Eventually, Brunner figured that to get any traction, he would personally have to become the machine's product manager. "It didn't come out of one of the product groups, and it wasn't going anywhere," said Brunner. "There was this process at Apple, where they decided if something was going to get on the product road map. There was a presentation that you needed to put together and a group you needed to present to. I acted like a product marketing guy and presented this idea and got it off the ground."

As Spartacus was finalized for market, it was discovered that the integrated speakers presented a major problem: When the volume was cranked up, the internal CD-ROM skipped. The skipping vexed the team for several months until an engineer from Bose suggested a solution. He recommended the use of a much smaller pair of speakers on the desktop and adding a subwoofer on the floor, which could also accommodate the machine's power brick. The fix worked, and the machine could deliver room-filling audio with only forty watts of power.

Making the changes meant a new working prototype wasn't ready until December 1995. Then it was decided to add a newer, updated circuit board, and a larger liquid crystal display (LCD) screen. In June 1996, Prototype Two finally emerged.

To Jony and the rest of the design team, who had lived with various prototypes for more than three years, the dark gray enclosure with mahogany trim had begun to look old. They had other doubts too. "Some of us felt the color was too strong," Jony recalled. "But we had all looked at the concept so much, we couldn't decide what color it should be."[31]

An outside color consultancy was hired. They came up with the brilliant suggestion that the designers shouldn't focus on the color of the computer; they should look at the color of the environments it would be put in. To find the right color, the consultants put together several palettes of cloth, wood, leather and carpeting, representing the colors found in a typical home. Several prototypes were painted and compared to each palette under different lighting conditions. A dozen options became three and, finally, just one, a metallic green/gold. Thanks to the metallic sheen, the bronze color had a chameleon-like effect that reflected the colors around it, helping it blend into any room. The mahogany accents were switched to black leather, which would likely wear better than wood.

Initially, the design department loved the final result. They thought it was a great all-purpose computer with good entertainment options in a high-end, high-quality proposition. This latest model included a TV/FM tuner, which allowed it to transform from computer to stereo to TV. Tim Parsey summarized the effects and qualities of the machine: "It's really complicated geometrically. But it doesn't look complicated. From the front, the design is quite simple. Yet it embraces the user in a powerful way. It's incredibly thin, yet the back tells you that it's strong enough to support itself with ease. And every curve and detail has a purpose."

Jony saw a deeper virtue, observing that "it challenges our perceptions in a fundamental way."

In August 1996, the third working prototype finally rolled off an actual production line, proving the machine could be made in quantity. In September, tooling was completed and the final design was finished in December 1996. It was more than four years since Brunner had written his conceptual brief.

With the much-anticipated twentieth anniversary of the Apple approaching, the decision was made to designate Spartacus as a special edition. Officially named the Twentieth Anniversary Macintosh, the new product was limited to a run of just twenty thousand units. Apple unveiled it at Macworld in January 1997 and the first two units were given to Steve Jobs and Steve Wozniak, who had just returned to the company as advisers.

To make it more memorable, the machine was hand-delivered to customers' homes by specially trained "concierges," who set up the machines, installed any expansion cards (along with the ugly hunchback) and showed users how to use them.

"I think it is the first sensible computer design that we have seen in a long time," said Henry Steiner, Hong Kong's most eminent graphic designer. "It is quite beautiful and desirable. It has the status value of a Porsche. The fact that the machine combines computer, television and stereo system is impressive."

Like the MessagePad, the Twentieth Anniversary Macintosh (TAM) won not only kudos but awards, including the Best of Category prize for *I.D.* magazine's Annual Design Review.

Steve Wozniak thought it was the perfect college machine "with the computer, TV, radio, CD player and more (AV even) all in one sleek machine." He had several at his mansion in the hills of Los Gatos above Silicon Valley. By the time the machine was pulled from the market one year after launch, however, Wozniak seemed to be the only person on the planet who liked it.

The TAM bombed in the marketplace. The machine widely missed its mark. Originally priced at $9,000, within a year the list dropped to under $2,000. It was originally intended as a mainstream product, but the marketing group turned it into a pricey special edition. It was the last straw. After all the battles to get the TAM to market, Brunner had grown tired of Apple's dysfunctional culture.

Bye-bye, Brunner

Just before the release of Twentieth Anniversary Mac, Brunner quit. He went to join Pentagram, the prestigious international design firm founded in London, which had courted him earlier that year.

Brunner's resignation had been in the cards for months. Early in 1996, he had taken an extended leave of absence. Though he returned that fall, things had gone from bad to worse within Apple and the design department. Everyone was frustrated. Two other longtime designers announced they were leaving before Brunner took his departure in December.

The decisive factor in his leaving was undoubtedly the Twentieth Anniversary Mac. After all the battles to get it to market, Brunner believed bungled positioning and pricing led to the machine's failure. "It was never intended to be a special edition thing," said Brunner. "It was intended to be higher-end but a mainstream product. . . . [I]t was a very provocative, forward-looking design, and it foretold what was coming six or seven years later."

More important, Brunner said, the process—fraught and hard fought as it had been—represented a line in the sand from the design group, an attempt to change Apple's internal culture. "That was when we as a design group said we are not going to be a service to other parts of the organization. We are going to take these ideas and push them forward on our own. It pissed people off but it also opened people's eyes up to what a truly design-driven process can do."

As the failure of the anniversary Mac indicated, Apple had become dysfunctional; it was a struggle to get products out, and there were constant battles with engineers and executives. "I quit for two reasons," explained Brunner. "One, the job wasn't fun, and to be brutally honest I was losing interest in it. I was spending more and more time in management meetings where I would be there for eight hours and only really needed to be there for thirty minutes. You feel like you are atrophying, you are wasting away. I'm not the kind of person that can just do the job even though you fucking hate it. Can't do it."

With Brunner's departure, Apple faced growing chaos. There was pressure once again to conduct a search for a name-brand designer, just like the company had done five years earlier. Brunner advised against looking outside for their next design leader. Most of the design team would depart, he warned, and besides, Apple already had a superstar. The job should go to his deputy, Jony Ive.

"He had quiet leadership qualities and he was super respected," said Brunner. "Not to put the other guys down at all, for me there was no other choice."

For some at Apple, Jony's age and inexperience were at issue. He was only twenty-nine, but Brunner recommended Jony because he admired his quiet commitment. "He was very consistent, very strong, and he was very ambitious," said Brunner. "Not in the wear-it-on-your-sleeve kind of ambition that many people have, but he was very strong and insistent. I'm going to do this, and I'm going to do it well."

Most of all, Jony had what Brunner called "the full spectrum mentality." He saw the big picture and the details.

"Jony is very much the craftsman," Brunner explained. "He loves the big picture but he also revels in the details, being in the factory and knowing exactly where every screw goes. . . . I just knew he had the qualities to be successful."

In other words, Jony had what it took to succeed in a corporate environment. He was willing to sit through the endless meetings and battle middle managers to get his designs made.

"It would have been a disaster if they had hired a headhunter and hired a guy that had a name and offered him a ton of money," Brunner said.

Jony got the job. "It was probably one of the better recommendations I ever made," Brunner said.

Jony inherited a legacy at Apple that would help him thrive. "The Brunner era (1990–95) was by far the most productive and interesting period in Apple's design history," Paul Kunkel would write in *AppleDesign*. "IDg became the most visible and prestigious corporate design group in the world, won more design awards than the rest of the computer industry combined and reached a level where further improvement meant using its own work as a yardstick rather than the competition's." A string of successful and groundbreaking products set the template for the future, including the PowerBook (which anticipated today's MacBooks); the Twentieth Anniversary Mac (the flat-screen iMac); and the Newton, which was a crude precursor to the iPhone and iPad.

Perhaps even more important, Brunner built the studio, hired great talent and set up the culture. "Bob did more than lay the foundations for Jony's design team at Apple—he built the castle," said Clive Grinyer. "After Bob, it was the first time that an in-house design team was cool."[32]

No design slouch himself, Brunner became a partner in the San Francisco office of Pentagram in 1996. He worked with Amazon on the original Kindle, and with Nike and Hewlett-Packard, among many others. In 2007, Brunner helped create the Beats by Dr. Dre brand of headphones, which have been a mega success. In mid-2007, Brunner founded Ammunition, a design consultancy in San Francisco, where he's worked with Barnes & Noble, Polaroid and Williams-Sonoma. He's won a ton of awards, and his work is included in the permanent collections of both the Museum of Modern Art in New York and the San Francisco Museum of Modern Art.

But Brunner likes to joke that the only thing he'll be remembered for is bringing Jony to Apple. "When I die, my tombstone is going to say: 'The Guy Who Hired Jonathan Ive!'"

Chaos Reigns

Brunner quit just in time—at least, for him. Just days after Jony took over, Apple warned that the crucial holiday buying season for 1995 would fall far short of expectations, thanks to an overabundance of cheaper, low-end systems, and a shortage of more profitable Power-Books and high-end desktops.

"Our warehouse was full of Yugos at a time when everyone was buying Mercedes," said Satjiv Chalil, the VP of marketing at the time.[33]

Until then Apple had appeared to be flying high. But that troubled holiday quarter would be followed by two years of plunging revenues, a free-falling stock price and a rotating door of lackluster CEOs. Apple's tumble was quick and dramatic. In 1994, Apple commanded nearly 10 percent of the worldwide multibillion-dollar market for personal computers, making it the second biggest computer manufacturer in the world after the giant IBM.

But in 1995, Microsoft released its new operating system, Windows 95, which took off like a rocket. Windows 95 was Microsoft's most shameful rip-off of the Mac operating system yet, but the software made their PCs good-enough facsimiles of the Mac. Cheap, utilitarian Windows 95 machines flew off the shelves, while Apple's overpriced, incompatible machines did not.

Microsoft licensed its operating system to dozens of hardware makers, who competed stiffly and drove down prices. To stay afloat, Apple tried a desperate tactic. It licensed the Macintosh operating system to several computer makers, including Power Computing, Motorola, Umax and others, but the Mac market remained flat.

In the first quarter of 1996, Apple reported a loss of $69 million and laid off 1,300 staff. In February, the board fired CEO Michael Spindler, who had taken over for John Sculley, appointing in his place Gil Amelio, a veteran of the chip industry with a reputation as a turnaround artist. But in the eighteen months that Amelio was on the job, he proved ineffectual and unpopular. Apple lost $1.6 billion, its market share plummeted from 10 to 3 percent, and the stock collapsed. Amelio laid off thousands of workers, but he was raking in about $7 million in salary and benefits while sitting on $26 million in stock, according to the *New York Times*. He lavishly refurbished Apple's executive offices and, it was soon revealed, negotiated a golden parachute worth about $7 million. The *Times* reported employees' view that Apple's governance during this period was a "kleptocracy."

Internally, the company was extremely fractured, split into dozens of different groups, each with its own agenda, which often conflicted. To make matters worse, Apple had become an experiment in extreme democracy. In reaction partly to the tyrannical ways of Steve Jobs, the company had transformed itself into a bottom-up, rather than a top-down organization.

There had to be consensus on every decision, involving all the interested parties. Steering committees would be set up to guide new products to market. As product designer Terry Christensen put it: "A lot of people considered Jobs' approach tyrannical and misguided. Funneling an entire project through one person, be it Jobs or another visionary leader, inevitably resulted in lopsided products that exhibited all the strengths and weaknesses of its creator, like the first Mac. Instead, the steering committee approach brought every discipline involved in a project together—engineering, software, marketing, product design, industrial design, manufacturing—and required discussion and consensus at every stage of development."[34]

Product development by consensus proved extremely bureaucratic. Whenever a new product was proposed, three documents had to be drawn up: a marketing requirement document, an engineering requirement document and a user-experience document. Mark Rolston, SVP of creative at Frog, summed it up this way: "Marketing is what people want; engineering is what we can do; user experience is 'Here's how people like to do things.'"[35]

The three documents would be sent upstairs to be reviewed by a committee of executives. If they were approved, a team leader (the "champion") would be assigned to the project and the design group would get a budget. Then it would go back to the marketing, engineering and user-experience groups for more work. Don Norman: "The team would work on expanding the three requirement documents, inserting plans on how they hoped to meet the marketing, engineering and user-experience needs—figures for the release date, ad cycle, pricing details and the like."[36]

Norman said in some ways, it was "a well-structured process" but he acknowledged its shortcomings. Not only was it slow, cumbersome and bureaucratic, it inevitably led to compromises. When one team wanted to do it like this, and another team like that, feature creep took over, resulting in a lack of cohesion in the product.

"The businessman wants to create something for everyone, which leads to products that are middle of the road," said Brunner. "It becomes about consensus, and that's why you rarely see the spark of genius."[37]

Even if a great idea came along, it was impossible to get anything done. Norman described just such an occasion.

"I remember vividly Jony Ive coming to me one day," said Norman. "Then, high-end Apple customers would buy a machine and the first thing they wanted to do was take it apart and add a lot more memory, or add a video card or a second processor. And it was a pain to open the machine and you couldn't access the memory and you had to take out some parts. And Jony had figured out a way to get around this, with a desktop computer that had two fasteners which you simply unhooked . . . making it trivial to add and change the machine's memory. And I thought this was marvelous.

"The problem was, the hardware people refused to build it. So Jony and I went around the company going from vice president to vice president trying to convince them. If they said the idea would be too expensive, Jony presented them a price analysis showing that it wasn't; if they said it would take too long to produce and ship the new idea, Jony would show them that he had already spoken to the factory, and they could do it in the amount of time available. It went on and on, for several weeks. Finally the CEO adopted Jony's idea and it was accepted. But that was the old Apple—it wouldn't happen like that today."

Jony's idea for easy access to the insides eventually made its way to market with a Power Mac 9600 in August 1997, and became a mainstay for all the Macintosh tower designs that followed.

Sometimes the fiefdoms, the bureaucracy, doomed solid new ideas. The most interesting product to come out of Jony's design group in the late 1990s was the eMate, a small, inexpensive plastic computer for schoolkids. It was a curvaceous clamshell made of translucent green plastic, whose see-through look would be copied for the first iMac. Everyone wanted one. It had a huge lust factor, but it bombed.

"This product was driven by industrial design and it died because it didn't have the right champion at Apple," Norman said. "It died because there was a fight over it among the different divisions. Should it run Newton software? Should it run Apple OS software? Which existing computer was it going to compete with? No one came forward and said, 'This could be a great computer for schools. What kind of software should it run, and how are kids going to use it?' No one looked into its use backwards, starting from what kids would want to do with it and so what should go into it. And so due to the lack of cohesion at Apple, the eMate died."

The eMate's uniqueness was a rarity in the interim that followed Brunner's departure. Most of the products were boring me-too designs, which reflected the engineering-driven culture at the core of the company. Much of their work was skin jobs. The chaos of trying to create in what seemed like an increasingly adversarial environment wore on Jony. Only a few months after being put in charge of the design group, he also was thinking of quitting.

"It was a company that certainly wasn't innovating," Jony said at the time. "We lost our identity and looked to competition for leadership."[38]

Amelio had little appreciation for design. "There wasn't that feeling of putting care into a product, because we were trying to maximize the money we made," Jony said. "All they wanted from us designers was a model of what something was supposed to look like on the outside, and then engineers would make it as cheap as possible. I was about to quit."[39]

Before Jony could quit, Jon Rubinstein, his new boss, talked him out of it. Just recruited as Apple's head of hardware (the same job he'd held working with Steve Jobs at NeXT), Rubinstein gave Jony a raise and told him, going forward, things would be different.

"We told him that we were going to struggle to get through where the company was then, and that once we turned the company around, we were going to make history. Those were the terms we used to keep him at Apple—and also that henceforth, design was going to be really valued at the company."[40]

Rubinstein's promise would be fulfilled. The era during which it took three years to get products out the door did end; in the coming years, the rate at which new products and new ideas were adopted—many of them from Jony Ive's fertile brain—would be nothing less than remarkable.

CHAPTER 5

Jobs Returns to Apple

The thing is, it's very easy to be different, but very difficult to be better.   —JONY IVE

On the morning of July 9, 1997, several dozen members of Apple's top staff were summoned to an early-morning meeting. In an auditorium at company HQ, Gilbert Amelio, who'd been Apple's CEO for approximately eighteen months, shuffled onto the stage. "Well, I'm sad to report that it's time for me to move on," he said, then quietly left the auditorium. Apple's board had just fired him.

Fred Anderson, the interim CEO, said a few words before Steve Jobs took the stage. Jobs had been brought in as an adviser when Apple bought NeXT, his struggling software company, and, after firing Amelio, the board asked him to take over.

Jobs looked like a bum, wearing shorts and sneakers and several days' stubble. It was almost exactly twelve years since he had been ousted from the company over a previous July 4th weekend.

"Tell me what's wrong with this place," he said to the group.

Before anyone could reply, he burst out: "It's the products. The products suck! There's no sex in them anymore."

Jony was in the room, sitting toward the back. He wanted to quit. But as he sat there thinking about returning to England with his wife, Jobs said something that gave him pause. Jobs told the group that Apple would be returning to its roots. "I remember very clearly Steve announcing that our goal is not just to make money but to make great products," Jony later recalled. "The decisions you make based on that philosophy are fundamentally different from the ones we had been making at Apple."[1]

Much was about to change in how Apple was run, beginning with the product lineup. When Jobs returned to Apple in 1997, the company had forty products on the market. To appreciate the baffling nature of Apple's kitchen-sink strategy at the time, consider the company's computer lineup.

There were four main lines: the Quadra, the Power Mac, the Performa and the PowerBook. Each was split into a dozen different models, which were delineated from one another with confusing product names straight out of a Sony catalog (for example, the Performa 5200CD, Performa 5210CD, Performa 5215CD and Performa 5220CD). And that was just computers. Apple had branched out into a wide-ranging product portfolio, selling everything from printers, scanners and monitors to Newton handhelds.

To Jobs, this made no sense.

"What I found when I got here was a zillion and one products," Jobs later said. "It was amazing. And I started to ask people, now why would I recommend a 3400 over a 4400? When should somebody jump up to a 6500, but not a 7300? And after three weeks, I couldn't figure this out. If I couldn't figure this out . . . how could our customers figure this out?"[2] The product line was so complicated that Apple had to print elaborate flowcharts to explain to customers (and as a cheat sheet for employees) what the differences between Apple's products were.

As chaotic as Apple's portfolio was, it had nothing on the anarchy of Apple's organization chart. Apple had grown into a bloated Fortune 500 company with thousands of engineers and even more managers, many of whom had overlapping jobs and responsibilities. Lots of them were exceptionally talented, but there was no central command and control.

"Apple, pre-Jobs, was brilliant, energetic, chaotic and nonfunctional," recalled Don Norman.

In fact, Apple's restructuring was under way before Jobs returned. Jobs joined the fray. He looked at everything: product design, marketing, the supply chain. Jobs started a thorough product review; he set up in a large conference room and called in the product teams, one at a time. The teams, often numbering twenty or thirty people, would present their products and take questions from Jobs and other executives. At first they wanted to give PowerPoint presentations, but Jobs quickly banned them. He saw PowerPoints as rambling and nonsensical; he preferred getting people to talk and asking them questions. In these meetings, it soon became clear to Jobs that Apple was a rudderless ship.

After several weeks, during a big strategy meeting, Jobs had had enough.

"Stop!" he screamed, "This is crazy."

He jumped up and went to the whiteboard. He drew a simple chart of Apple's annual revenues. The chart showed the sharp decline, from $12 billion a year to $10 billion, and then $7 billion. Jobs explained that Apple couldn't be a profitable $12 billion company, or a profitable $10 billion company, but it could be a profitable $6 billion company.

That meant radically simplifying Apple's product pipeline. How? Jobs erased the whiteboard and drew a very simple two-by-two grid in its place. Across the top he wrote "Consumer" and "Professional," and down the side, "Portable" and "Desktop."

Welcome to Apple's new product strategy, he said. Apple would sell only four machines. Two would be notebooks, the other two desktops. Two machines aimed at pros, two machines aimed at consumers.

It was a radical move, cutting the company to the bone. Under Amelio, the plan had been to offer more and more products. Jobs proposed the opposite. In a single stroke, Jobs doomed dozens of software projects, and eliminated almost every product from Apple's hardware lineup. Over the next eighteen months, more than 4,200 full-time staff were laid off. By 1998, Apple had shrunk to only 6,658 employees, half the 13,191 the company had in 1995.[3] But the balance sheet was brought back into control.

The most controversial decision of Jobs's first months was the late 1997 killing of the Newton, Apple's PDA, which, after Jony's Lindy, was in its seventh generation. A money loser from the start, Amelio's administration had tried to spin off the Newton into its own division but the then-CEO had changed his mind at the last minute. As an adviser, Jobs had tried to persuade Amelio to shut down the Newton. It had never really worked right, and it had a stylus, which Jobs hated. Despite a small and dedicated following, it hadn't taken off with a mass audience. Plus, Jobs saw it as John Sculley's baby. Though it was the only really innovative thing Sculley achieved under his tenure, Jobs had many reasons to end the Newton's brief life.

Most executives would have thought twice about killing a well-loved product, and Newton lovers flooded Infinite Loop's parking lots with placards and loudspeakers. ("I give a fig for the Newton," one sign read.) PDAs were on the rise, thanks to the success of handhelds like the Palm Pilot, but to Jobs, the Newton was a distraction. He wanted Apple to concentrate on computers, its core product.

Jobs aimed at making innovative products again, but he didn't want to compete in the broader market for personal computers, which was dominated by companies making generic machines for Microsoft's Windows operating system. These companies competed on price, not features or ease of use. Jobs figured theirs was a race to the bottom.

Instead, he argued, there was no reason that well-designed, well-made computers couldn't command the same market share and margins as a luxury automobile. A BMW might get you to where you are going in the same way as a Chevy that costs half the price, but there will always be those who will pay for the better ride in the sexier car. Rather than competing with commodity PC makers like Dell, Compaq and Gateway, why not make only first-class products with high margins so that Apple could continue to develop even better first-class products? The company could make much bigger profits from selling a $3,000 machine rather than a $500 machine, even if they sold fewer of them. Why not, then, just concentrate on making the best $3,000 machines around?

The potential merits of Jobs's strategy for the company finances were clear. Fewer products meant less inventory, which could have an immediate effect on the bottom line. In fact, Jobs was able to save Apple $300 million in inventory in just one year, and avoid having warehouses full of unsold machines that might have needed to be written off if they failed to sell.

Inventing Steve Jobs, 1976 and After

Jobs's plan for Apple was more than a matter of B-school economics: He planned to make industrial design the centerpiece of Apple's comeback. Since his first incarnation at Apple (1976–1985), it had been apparent that design was a guiding force in the trajectory of Steve Jobs's life.

Unlike Jony, Jobs had no formal design training, but he possessed an intuitive design sense that dated to his childhood. Jobs, early on, learned that good design wasn't just on the exterior of an object. As Mike Ive had been for Jony, Jobs's father was a formative influence on his son's appreciation of design. "[My father] loved doing things right. He even cared about the look of the parts you couldn't see," Jobs recalled. His father refused to build a fence that wasn't constructed as well on the back side as it was the front. "For you to sleep well at night, the aesthetic, the quality, has to be carried all the way through."[4]

Jobs grew up in a house inspired by the tract homes of Joseph Eichler, a postwar developer who brought a mid-century modern aesthetic to the architectural landscape of California. Although Jobs's childhood home was probably a knockoff of an Eichler (what Eichler fans call a "Likeler"), it left an impression. Describing his childhood home, Jobs said, "I love it when you can bring really great design and simple capability to something that doesn't cost much. It was the original vision for Apple."[5]

For Jobs, design amounted to more than appearances. "Most people make the mistake of thinking design is what it looks like," Jobs famously said. "People think it's this veneer—that the designers are handed this box and told, 'Make it look good!' That's not what we think design is. It's not just what it looks like and feels like. Design is how it works."[6]

With the development of the Macintosh, Jobs got really serious about how-it-works industrial design, which he believed could be a key differentiator between Apple's consumer-friendly, works-right-out-of-the-box philosophy and the bare bones, utilitarian packaging of early rivals like International Business Machines.

In 1981, with the PC revolution not yet five years old, 3 percent of U.S. households had a personal computer (including toy systems like the Commodores and Ataris). Only about 6 percent of Americans had even encountered a PC at home or work. Jobs understood that the home market presented a huge opportunity. "IBM has it all wrong," he would say. "They sell personal computers as data-processing machines, not as tools for individuals."[7]

Jobs and his chief designer, Jerry Manock, went to work on the Mac, with three design constraints. To keep it cheap and make it easy to manufacture, Jobs insisted on just one configuration, an echo of his hero Henry Ford's Model T. Jobs's new machine had to be a "crankless computer": A new owner should just be able to plug the machine into the wall, press a button and it would work. The Macintosh would be the world's first all-in-one PC, with the screen, disk drives and circuitry all housed in the same case, with a detachable keyboard and mouse that plugged in the back. In addition, it shouldn't take up too much space on a desk, so Jobs and his design team decided it should have an unusual vertical orientation, with the disk drive below the monitor, instead of to the side like other machines at the time.

The design process continued for several months, with a sequence of prototypes and endless discussions. Material evaluations led to the use of tough ABS plastic that was used to make LEGO bricks, which would give the new machine a fine, scratch-resistant texture. Troubled by the way earlier Apple IIs had turned orange in sunlight over time, Manock decided to make the Macintosh beige, initiating a trend that would last twenty years.

As Jony would do in the next generation at Apple, Jobs paid close attention to every detail. Even the mouse was designed to reflect the shape of the computer, with the same proportions, and a single square button that corresponded to the shape and placement of the screen. The power switch was put around the back to stop it being switched off accidentally (especially by curious kids), and Manock thoughtfully put a smooth area around the switch to make it easier to find by touch. "That's the kind of detail that turns an ordinary product into an artifact," Manock said.[8]

The Macintosh looked like a face, with a slot for the disk drive resembling a mouth and a chinlike keyboard recess at the bottom. Jobs loved it. This is what made the Macintosh look "friendly"—an anthropomorphic smiley face. "Even though Steve didn't draw any of the lines, his ideas and inspiration made the design what it is," designer

Terry Oyama said later. "To be honest, we didn't know what it meant for a computer to be 'friendly' until Steve told us."[9]

It took five years—the Macintosh, conceived in 1979, was released in January 1984—but the product represented the first distillation of Jobs's design philosophy. Unfortunately, the Macintosh was the last product that Steve Jobs would see to market during his first tenure at Apple. About eighteen months after launching the Mac, in September 1985, Steve Jobs lost a boardroom power struggle. John Sculley, the ex-PepsiCo marketing executive Jobs himself had recruited, took over. His design philosophy, though, would continue to be influential throughout his absence.

Before his departure, Jobs talked about making Apple in the eighties what Italy's Olivetti had been in the seventies, the undisputed world champion of industrial design. Design was hot in the eighties, especially in Europe, with groups like Italy's Memphis Group[10] of architects and designers earning accolades for their bold, colorful designs (memorably described as "a shotgun wedding between Bauhaus and Fisher-Price"[11]). In March 1982, two years before the Macintosh was unveiled, Jobs decided Apple needed a world-class industrial designer to craft a uniform design language for all the company's products.

At the time, Apple's hardware was all over the place. The company's different divisions—the Apple II division, the Mac division, Lisa peripherals—were all using different designers with different ideas. Apple's products looked like they came from four different companies, not one. It drove Jobs crazy.

Jobs had Manock set up a design competition, in which potential candidates were asked to draft seven products, each named after a dwarf from Snow White. The name was inspired by the storybook Manock was reading to his young daughter; Jobs loved that it conjured up images of products that were distinctive, friendly and with personality.

Almost from the start, the front-runner was Hartmut Esslinger, a German industrial designer then in his mid-thirties. Like Jobs, Esslinger was a college dropout. He had gained notice designing TVs and other consumer electronics for Sony and Wega, a German company that Sony eventually bought. One of his TV designs for Wega was in bright green plastic, which the CEO of the company nicknamed "frog." This no doubt influenced Esslinger to name his company Frog Design, which is also an acronym for Esslinger's homeland: Federal Republic of Germany.

In May 1982, Esslinger flew to Cupertino to meet Jobs. They were alike—both natural entrepreneurs, brash and opinionated—and they bonded over a love of Braun and Mercedes. Jobs was particularly impressed that Esslinger had worked for Sony, a design-centric company that Jobs wanted Apple to emulate.

A master at pitching his ideas and philosophy, Esslinger also knew how to work hard. His group labored through four major design phases, and after months of work, put on an overwhelming show for Apple's brass. While the other contenders in the competition made a handful of models, Esslinger's group turned out forty beautifully finished models, two or three variations for each product. The other groups pitched designs in dark plastic with hard edges (like Sony's stereo components from the eighties), but Esslinger's designs were simple and sophisticated, made from lightly textured, cream-colored plastic. Like Jony, Esslinger wanted to differentiate Apple from the masculine design of eighties electronics and create a design language based on recurring elements that echoed the consistency of software that the Mac already provided.

Jobs was delighted with the formal presentation of Esslinger's work in March 1983: Esslinger was declared the winner of Snow White and soon emigrated to California to set up his own studio, Frog Design, Inc., having agreed to provide exclusive services to Apple for an unprecedented $100,000 a month, plus billable time and expenses. Billings would quickly add up to $2 million a year, far more than competing design firms were earning from their clients.

Jobs unceremoniously told Manock and the other in-house designers that they would be working for Esslinger, who was essentially an outside contractor, albeit one with special status. Manock was killing himself designing the first Mac, but Jobs told the hapless designers that they should consider themselves lucky to be in a position to learn from the talented Esslinger. Most worried about their tenuous jobs and, indeed, the move more or less ended Manock's career at Apple.

Esslinger's emerging Snow White style made Apple's best design efforts look clumsy and outdated. Snow White designs would eventually win all the major industry awards and be so widely copied that it became the de facto design language for the entire PC industry. It was too late to redesign the Mac, at that point, which had already been tooled at great expense, so Frog's first major product for Apple using Snow White would be the Apple IIc, the fourth in the line of Apple IIs and the first attempt at a portable computer (the "c" stood for "compact"). More important, it was Apple's first design-driven product (designed from the outside in, rather than the inside out); even with Jobs gone, he left behind a design evolution as his legacy.

That shift in emphasis was compelling beyond the design department: Apple's engineers bent over backward to accommodate Esslinger's designers, rather than fighting their ideas. It was a small but subtle shift; an early attempt to make Apple design driven rather engineering driven. By the time he returned, however, the old paradigm, with the engineers wielding the power, had returned, as Jony Ive had discovered.

The "A" Team

Jobs's plan for switching up the teams at Apple upon his return was just as straightforward as his notions for simplifying the product portfolio.

He would cut back so that his "A" team—the company's best designers, engineers, programmers and marketers—could concentrate on making innovative products.

Jobs already trusted his old NeXT executives, but he looked to spot existing Apple talent and elevate them in the ranks. The people survey he conducted led to a streamlining of Apple's organizational chart. Jobs insisted on a clear chain of command all the way down the line. Everyone in the company knew to whom they reported and what was expected of them.

As Jobs told *BusinessWeek*, "Everything just got simpler. That's been one of my mantras—focus and simplicity."[12]

During the process of his product and personnel reviews, Jobs called a meeting with half a dozen top analysts and journalists covering Apple. He wanted to explain to them his new game plan.

"He specifically emphasized getting back to meeting the needs of their core customers and said that Apple had lost ground in the market because they were trying to be everything to everybody instead of focusing on the real needs of their customers," said Tim Bajarin, an analyst with Creative Strategies, who was in the room. "He also pointed out that Apple had broken new ground with the original Mac OS and hardware designs and that he would now make industrial design a key part of Apple's strategy going forward."[13]

Bajarin, for one, was skeptical.

"My first impression was that Apple had so many problems that I could not see how industrial design needed to be a key part of his strategy to save Apple," Bajarin recalled. "I also was concerned that given Apple's serious financial situation, whatever he was going to do needed to be rock solid and have an impact quickly."

Bajarin had another recollection: "I also remember telling the people I was with that you can never underestimate Steve Jobs and that if anybody can save Apple, it would be Jobs."[14]

Despite his talk about returning Apple to a design-led company, Jobs didn't immediately visit the ID studio. Brunner's strategy of putting the studio off campus almost backfired, because, unaware of what he already had, Jobs went to look for a world-class designer from outside the company.

He thought seriously about bringing back his old design partner, Hartmut Esslinger of Frog Design, who had been working with NeXT. He called on Richard Sapper, who did the IBM ThinkPad laptop, and, incredibly, the car designer Giorgetto Giugiaro, whose run with Apple a few years before had produced nothing. Jobs also considered the famous Italian architect and designer Ettore Sotsass, who had catapulted Olivetti to the forefront of ID in the sixties.[15]

Across the road, Jony Ive realized his team was in jeopardy and that he had to demonstrate to his new boss what his shop could do. He put together brochures showcasing their best design work. He included in the glossy pamphlets concepts for evolving Apple's design language, which, with the translucent eMate, had just started moving in a bold new direction. "We generated small booklets to represent the team's capabilities," said a former designer on the team. "I think that played a great role in how Steve Jobs coming back perceived the team and its capabilities."

When Jobs finally took a tour of Apple's design studio, he was bowled over by the creativity and rigor he saw. The studio was full of eye-catching mock-ups that the previous regime had been too timid to consider. Jobs also couldn't help but notice the computer numerical control (CNC) milling machines and a fledgling computer-aided design (CAD) group.

Mostly, though, he bonded with the soft-spoken Jony, who would later say that he and Jobs saw eye to eye immediately. "We discussed approaches to forms and materials," Jony recalled. "We were on the same wavelength. I suddenly understood why I loved the company."

Jobs decided to keep the ID group intact, with Jony in charge. He initially made Jony report to Jon Rubinstein, head of hardware. (Eventually that would change, when design became its own autonomous division.) Even though Jony reported to Rubinstein, over the next few months Jony and Jobs began having lunch together. Jobs visited the Valley Green design studio too, often dropping by at the end of the day.[16]

"He would come over all the time," said a former member of the design team. "He'd come mostly to see Jony, but also to see what was going on." In time, Jobs become almost a fixture there.

Jobs Brings Focus

When Jobs returned, he brought focus not just to the company but to Jony's design group.

Jony was ostensibly in charge of the design squad, but the team's efforts weren't unified. Young and inexperienced as a manager, Jony wasn't exerting much discipline or leadership. It was creative chaos, just like the rest of the company. The design department was full of talented but willful designers, each working on his or her own projects with little or no coordination.

"Each designer had his own agenda, or their own design impulse, and there was no control [over their activities]," designer Doug Satzger said. "One designer had an agenda over what a laptop should be; another had an agenda over what a printer should be. There was no consistency on what the next Power Mac tower should be. The design group was not set up so that designers could work collaboratively as a team. All the designers were independent and had their own strong design sense. It was like they were all working for different companies."

Satzger said three designers were working on three different updates of the Power Mac, a powerful computer for professionals that came in a tower configuration. "There was no consistency," said Satzger. "Danny [De Iuliis] had designed a perfect cube with wheels on it. It was pretty big. Danny Coster was working on a model that consisted of various blocks thrown together, while the design of Thomas Meyerhoffer's [another member of the ID team] for the tower was all slurpy with lines all over—it was a monolithic piece of art. And the team was going ahead with not one of them but all of them."[17]

Esslinger and Brunner, in their times, had allowed designers to explore different design directions. Brunner in particular liked to treat the early design process almost as a competition. But he'd eventually choose the best ideas, and Esslinger, too, would unify the best ideas into a single design direction. Jony had worked in this kind of environment for several years, but didn't seem willing or able to provide a strong lead.

Jobs stepped in, shutting down unpromising projects and trimming back Apple's product line to suit his 2×2 matrix. Satzger remembers Jobs coming to the studio and telling the designers that Apple would be putting all its energies behind just four products. First and foremost would be a desktop for consumers. "Steve said, 'My daughter is going to college, and I've looked at everything out there and they're all crap. There's a real opportunity. Our target now is to build an Internet computer.' He was envisioning the iMac. That was the new focus."

Jobs wanted an inexpensive computer, something that would appeal to mainstream consumers eager to try out the Internet, which was just becoming popular thanks to Netscape's Navigator browser, cheap modems and an explosion in Internet service providers (ISPs) like AOL, which offered inexpensive Internet access plans. And he wanted it quickly. He'd cut back the company to give it breathing room, but he needed new products quickly to restore plunging sales. He was betting the company on this one product.

At the time, Apple's cheapest computer was $2,000, more than $800 above the average Windows PC. To be competitive against cheaper Windows offerings, Jobs initially pushed for a radically stripped-back machine called a "network computer" (also known as an NC), a hot idea in Silicon Valley at the time. The NC would be a cheap, simple terminal that connected to a central server over the Internet. It had no hard drive or optical disk drives, just a screen and keyboard. It was perfect for schools and workplaces and it seemed, at first glance, ideal for consumers eager to access the Internet.

Before Jobs's return, in May 1996, Apple had joined Oracle Corporation and thirty other hardware and software companies in the Networking Computing Alliance, which set the standard for cheap, diskless computers based on a common networking platform. Jobs's billionaire best friend, Larry Ellison, was especially bullish on NCs as the future of the computer industry. And as a newly installed member of Apple's board, Ellison told the press that Apple was building an NC. He'd recently launched a start-up, Network Computing Inc., to kick-start the sector.

Influenced by Ellison's thinking, but also eager to compete with him, Jobs also talked up the NC idea. "We're going to beat Ellison at his own game," he told his Apple colleagues with relish.[18] Just as he'd done with the first Macintosh, Jobs began by laying out certain specifications: The Mac NC should be an all-in-one product, ready to use right out of the box, in a distinctive design that made a brand statement. And it should sell for $1,200 or so. "He told us to go back to the roots of the original 1984 Macintosh, an all-in-one consumer appliance," recalled Apple's marketing chief, Phil Schiller. "That meant design and engineering had to work together."[19]

In September 1997, Jon Rubinstein instructed Jony to take Jobs's Mac NC specs and prepare a dozen foam models of potential designs.

Jony gathered the entire team in the design studio to hash it out. They started by discussing the Mac NC's potential target market. "We didn't start with engineering dictates," Jony said. "We actually started with people."

"The iMac revolved not around chip speed or market share but squishy questions like 'How do we want people to feel about it?' and 'What part of our minds should it occupy?'" Jony said later in a *Newsweek* interview.[20]

Jony was looking for the Mac NC's "design story." As his dad, Mike, had instilled in him, developing the design story was an essential first step in conceiving something entirely new. "As industrial designers we no longer design objects," Jony said. "We design the user's perceptions of what those objects are, as well as the meaning that accrues from their physical existence, their function and the sense of possibility they offer."[21]

They discussed topics like "objects that dispense positive emotions"; one of the designers suggested a transparent gumball dispenser as an example of this. The IDg also discussed how other businesses, like the fashion industry, might approach the problem. "We talked about companies like Swatch—companies that broke the rules—that viewed technology as a way to the consumer, not the consumer as the path to the technology," Jony said.

Later, Jony explained his thinking this way. The computer industry "is an industry that has become incredibly conservative from a design perspective," he said. "It is an industry where there is an obsession about product attributes that you can measure empirically. How fast is it? How big is the hard drive? How fast is the CD? That is a very comfortable space to compete in because you can say eight is better than six."[22]

But Jony offered a key insight: "It's also very inhuman and very cold. Because of the industry's obsession with absolutes, there has been a tendency to ignore product attributes that are difficult to measure or talk about. In that sense, the industry has missed out on the more emotive, less tangible product attributes. But to me, that is why I bought an Apple computer in the first place. That is why I came to work for Apple. It's because I've always sensed that Apple had a desire to do more than the bare minimum. It wasn't just going to do what was functionally and empirically necessary. In the early stuff, I got a sense that care was taken even on details, hard and soft, that people may never discover."[23]

Sitting around a table, Jony and the rest of team started sketching. Satzger remembered the table was covered with loose sheets of copy paper, colored pencils and Pilot pens. The team drew up ideas together as a collective, passing the loose sheets between them as they worked. Trying to imagine a machine that dispensed positive emotions, as they hoped the iMac would, designer Chris Stringer made a beautiful sketch of a colorful candy dispenser.

Satzger remembers coming up with the initial egg shape for the iMac, based on TVs he had previously designed for Thomson Consumer Electronics. "If you look at the shape, it's almost exactly the same profile."

The idea appealed to Jony and the rest of the team. Immediately, the decision was made to pursue the egg shape as the primary design direction.

The iMac had to be on the market in a matter of months or Apple would go out of business. To speed up the design process, Jony instigated a radical, integrated design process that transformed the way Apple developed its products. The workflow that the design group uses today is basically unchanged from the system Jony introduced.

The IDg needed new tools that could streamline the complex design process and allow the designers to create 3-D designs that, in turn, could be used by outside toolmakers at the factories to create the molds for the computer's case. Marj Andresen, who worked in the product design group, was instrumental in helping Jony find sophisticated CAD software to bring the new machine to life. The new CAD modeling process Jony helped pioneer integrated disparate computing systems, using complex software that translated files to make them interoperable.

"I was given nine months to get the tools into production," recalled Andresen, who calls herself a "CAD/CAM therapist" because of her role supporting high-strung designers. "Nine months to get from design to production was a very short time and impossible with drawings," said Andresen. "Our only hope was to use the design files directly. While the tools existed, the process hadn't yet been proved on either the engineering or the tooling end. It was crazy and hectic and exciting."

Prior to the iMac, hardware engineering (electrical design) and product design (mechanical engineering) drove the design process. "They designed the size of the enclosure due to engineering constraints and ID were tasked to develop 'skins' to go around this enclosure," said Paul Dunn, a former CAD manager at Apple. "When Jobs returned to Apple, he and Jony turned this process on its head."

Although Jony's group had a small CAD team in the studio, it was still the early days of computer-aided design. The designers worked mostly with hand-drawn sketches and some early, relatively primitive 2-D CAD software. But Jony's team needed to design in three dimensions, not two.

They found the answer in Alias Wavefront, a 3-D graphics package used in the aerospace, automotive and the fledgling computer-animation industry. Steve Jobs's other company, Pixar, had used it for some special effects in *Toy Story*, released in 1995.

"Apple was producing designs much more complicated than any of our computer rivals," said Dunn. "The surfaces on [the iMac] were more akin to the aerospace and automotive industries than the computer industry. . . . We were pushing the envelope."[24]

Alias Wavefront is, in effect, a sculpting tool; it defines the outer envelope of a product, much like a sculptor models a work in clay. As soon as the designers had a promising shape sketched out, they took it to the CAD group, who are also known as "surfacing guys." At the time, IDg had a small group of operators; it's now grown to about fifteen surfacing guys who are based inside the IDg studio. They run Alias (now called Autodesk Alias) on high-end Apple computers and Hewlett-Packard workstations. With Jony's designers watching over their shoulders, the CAD operators created outlines of the proposed designs. The idea was to make sure the basic shapes and scale were sound.

The process often takes several days. When the designers and surfacing guys settle on a reasonably good shape, they send the file to one of the studio's CNC milling machines to create a physical model. The initial models are cut from foam, and later, more detailed "hard models" are cut from acrylonitrile butadiene styrene (ABS) or RenShape, a dense red foam that cuts like wood and is good for surfaces.

The design group in Jony's early days also had an early and very expensive 3-D printer in the studio. "Apple had been at the forefront of modeling technology for many years," said Dunn. "From the early nineties, Apple's modeling group had a stereolithography (SLA) machine which could create complex 3-D models in several hours. The chemicals were extremely toxic, but the results were worth all the hassles."

As Jony discovered in college, making detailed models was a key part of the design process. "When you see the most dramatic shift is when you transition from an abstract idea to a slightly more material conversation," Jony said. "But when you made a 3-D model, however crude, you bring form to a nebulous idea, and everything changes—the entire process shifts. It galvanizes and brings focus from a broad group of people. It's a remarkable process."[25]

With this new workflow arrangement, Jony consolidated the model shop into the design studio (previously it had been under the aegis of product design). Andresen says it was a sensible move, organizationally and operationally. "The model shop gave the ID guys a first look and feel at the product design. They were creating one-off models of what products might look like. . . . [T]hey really had the coolest stuff."[26]

The modeling guys were essential to the product process. "Our model shop guys were craftsmen," Andresen remembered. "They could build anything, but they had to learn how to use the new computer tools and to accept the design files from the engineers."

Another important piece of software in reimagining the design process was Unigraphics, a 3-D engineering program developed by McDonnell Douglas for use in aerospace. Andresen and her group created software that allowed the 3-D models created by the surfacing guys in Alias to be imported into Unigraphics. From there, the product design group used Unigraphics to create workable products from the surfaces determined by the designers. The engineers would bring in detailed 3-D models of the product's components, allowing them to see whether the components would fit, and if the proposed shape would work. "The designers would sit with the CAD guys and say, 'Bring in the CRT tube. Here we need some volume for board, a place for connectors,' and so on," said Satzger. "Processes and interfaces were developed which allowed us to take these surfaces and import them into Unigraphics, which Product Design then used as the starting point to develop real solids and parts for manufacture," said Dunn.

The whole process was iterative. The designers worked constantly alongside the CAD operators and engineers, fiddling with the arrangements until they found a workable combination of outer skin and internal components. "It all sounds so simple, but it was very difficult, repetitive and time consuming," said Roy Askeland, a former Apple designer who worked on the iMac. "It was defining minute details, millimeters at a time."[27]

The last part of the puzzle involved sending the detailed 3-D CAD files to the mold makers. When it came to making the molds at the factory, toolmakers had previously relied on hand-drawn sketches and 2-D CAD printouts, which looked like blueprints. Although the factories had their own CAM systems to make the molds, they weren't always compatible with the designers' systems, so translating the design group's sketches and models into molds at the factory was still semi-manual and could take many months. The software Andresen helped create enabled all the systems to share common files, vastly simplifying and speeding up the process.

Although the parallel design workflow was overall faster with these improvements, the system didn't run smoothly. "It was a difficult move to pull off and it didn't feed nicely to CAM [computer-aided manufacturing] systems, either in house or at suppliers," said Andresen. "As we moved into surface modeling the limitations on the software and the hardware slowed us down considerably. Computers went from being boxy to having rounded corners and edges. These were very difficult to model and also very difficult for our suppliers to receive. . . . [W]e had multiple systems and even more system translation complexity."

Nonetheless, the bugs were worked out and the revolutionary process underlies IDg's workflow even today. "The design process basically revolves around two main systems: Wavefront and Unigraphics," said Dunn. "Yes, many interfaces and post-processors had to be developed, and some of them were very complicated, but the overall process wasn't that bad once it had been defined."[28]

Andresen said Apple's innovative designs of the late 1990s and early 2000s pushed the envelope of what CAD tools could handle at the time. "It's funny to think about now when I have 3-D modeling running on my iPad . . . but back then the idea of modeling what it looked like to build a computer out of aluminum and actually shine a light on the casing was unimaginable. The CAD vendors were struggling to keep up with our demands. We had to bring them into our ID and PD labs to show them what we needed. Between computers and automotive we were forcing the CAD industry to come up with lots of new software."

During that first month, Jony's team built at least ten models for the Mac NC—but they didn't show them all to Jobs. Jony's team knew right away they wanted the egg shape, so they shared only variations of that shape. "When it came to showing models to Steve Jobs, we would select models that we ourselves thought were good," said Satzger. "Steve would then approve—or not. Many times he said simply no to a model design. But we never presented anything to Steve that we didn't want him to pick."

When Jony first showed Jobs the egg-shaped machine in the studio, he rejected it. But Jony persisted. He agreed with Jobs that it wasn't quite right, but suggested it had a sense of fun. It was playful. "It has a sense that it's just arrived on your desktop or it's just about to hop off and go somewhere," he told Jobs.[29]

The next time Jony showed the egg shape to Jobs, the boss went nuts for it. Jobs started carrying the Foam Core model around campus, showing it to people to gauge their reaction. At the same time, he was cooling to the idea of a stripped-down network computer. Competing NCs already on the market, such as Microsoft's WebTV and Apple's own Pippin, sold by Bandai as @Mark in Japan, were getting zero traction in the marketplace.

Jobs ordered the NC be upgraded to a real computer with, among other things, a bigger hard drive and an optical drive. To prevent any delays, Rubinstein suggested the hardware should be based on the G3 desktop, a machine for professionals that had been in the pipeline before Jobs took over and had just been released. The addition of a hard drive and an optical drive meant enlarging the egg-shaped enclosure, but it was a relatively simple operation of scaling it up. Jony charged Danny Coster to take lead of the design.

The iMac was going to be made in plastic—it would be an "unashamedly plastic" product, in Jony's words—but plastic came with difficulties. "We didn't want it to look trashy," Jony explained later. "There's a fine line you walk between affordable and cheap. We certainly wanted it to appear affordable, and we wanted to make it clear that this isn't a terrifying technology—a technology that still alienates a huge number of people."[30]

Jony's group made models in a reddish blue, almost purple, and in orange, but the solid plastic looked cheap, so someone suggested the case be transparent. Jony immediately approved of the idea. "It came across as cheeky," he said. "That's why we liked translucency."[31] Transparent plastic had already started to creep into some of Apple's products, like printer trays and covers; the clamshell-shaped eMate, designed by Thomas Meyerhoffer, was made entirely of transparent plastic. At the time, Meyerhoffer told Macweek that translucency gave the eMate a sense of accessibility by allowing users to see inside.[32] Once Jony decided to make the iMac transparent, he realized the internals would have to be designed with care, too, since they would now be visible. Jony was particularly worried about the electromagnetic shielding that went around some of the internal components, which in opaque products was usually hidden in a big, ugly sheet metal box.

Jony had the designers bring all the different transparent-colored items they could find into the studio for inspiration. "We had a taillight from a BMW," said Satzger. "A lot of kitchen accessories. An old transparent thermos. Cheap flatware for picnics. We had a whole product shelf, full of these transparent products. We studied the qualities, the depth of a transparent product. The textures on the inside. The thermos was a big inspiration. It was a deep shiny blue, with a shiny thermos flask reflecting inside."[33]

Indeed, the final iMac, with its silver internal shielding visible through the transparent shell, resembles the transparent thermos married to a car taillight.

One of the designers brought in a small piece of beach glass, greenish blue in color, with a softly frosted surface. It might have come from Half Moon Bay, California, or Sydney's Bondi Beach in Australia, where Coster liked to surf. When Jony presented Steve with three models to choose among, he included this green/blue tint (that Jony's team called Bondi Blue) along with orange and purple models. Jobs selected the Bondi Blue.

The transparency added a sense of accessibility, but in order to give the iMac an even more approachable feel, the designers added a handle on top. For Jony, the handle on the iMac was not really for carrying it around, but to build a bond with the consumer by encouraging them to touch it. It was an important but almost intangible innovation that would change the way people interacted with computers.

"Back then, people weren't comfortable with technology," Jony explained. "If you're scared of something, then you won't touch it. I could see my mum being scared to touch it. So I thought, if there's this handle on it, it makes a relationship possible. It's approachable. It's intuitive. It gives you permission to touch. It gives a sense of its deference to you."[34]

The idea came with a downside, according to Jony. "Unfortunately, manufacturing a recessed handle costs a lot of money. At the old Apple, I would have lost the argument. What was really great about Steve is that he saw it and said, 'That's cool!' I didn't explain all the thinking, but he intuitively got it. He just knew that it was part of the iMac's friendliness and playfulness."

The iMac was code-named Columbus because it represented a new world. As Jony later expressed it, "With the first iMac the goal wasn't to look different, but to build the best integrated consumer computer we could. If as a consequence the shape is different, then that's how it is. The thing is, it's very easy to be different, but very difficult to be better."

Over the years, Apple had acquired a lot of legacy technologies, those elements that are becoming obsolete but aren't quite there yet. The fast-moving technology industry is full of them. At the time, Apple's legacy technologies included various parallel and serial ports for connecting mice, keyboards, printers and other peripherals. Like most computer makers, Apple tended to accommodate as many legacy technologies as possible. The company was loathe to lose a sale for the lack of a connector to hook up some old printer.

Jobs decided to make the iMac the first "legacy-free" computer. He ditched the old ADB, SCSI and serial ports, and included only Ethernet, infrared and USB. He also abandoned the floppy drive, a decision that drew more controversy than any other. These changes reflected Jobs's simplification philosophy, which would soon come into play in many products. Jony, too, would become a master of the approach, agreeing with Jobs's mantra: "Simplicity is the ultimate sophistication."[35]

USB, now a standard technology for connecting peripherals, was a particularly gutsy choice to include in the nineties. Invented by Intel, the USB 1.1 standard hadn't even been finalized yet (and wouldn't be until September 1998, a month after the iMac's release). It had no traction in the industry yet but Jobs was betting it would solve the increasingly vexing problem of special accessories for the Mac platform. As the Mac's market share shrank, fewer and fewer peripheral makers were willing to make hardware with special connectors just for the tiny Mac market. But USB peripherals could be made Mac-compatible with the addition of just the software to drive them, and, if necessary, Apple could write these so-called drivers itself.

To make the ports accessible, Jony located them on the side of the machine. "One of the things that makes the backs of most computers look so agricultural is the ton of cables pouring out," said Jony. "We moved the connectors to the side, which, functionally, makes them much easier to get to and keeps the back quite simple. The back of my computer may as well be its front in terms of what you see."

Characteristically, Jony sweated the details, including the power cable, which he also wanted to be transparent. "You know how, when you take a shower, condensation forms on the glass? We wanted that same kind of exquisite matte surface finish on the cable."[36]

The transparent mouse made Jony particularly proud. "If you know how mice work, it's quite intriguing," he said. "You see through the Apple logo, like a little window on the top of the mouse, into this little mouse factory. You see the ball moving on twin axles—well, it's actually pretty complicated and intriguing what goes on inside the mouse besides the ball rolling. We've tried fairly hard to layer what you can see inside. For the most part you just get a sense of what's inside—a sense of materials reflecting light and a sense of forms and shapes. It's only occasionally that you get a more literal view of what's going on inside."[37]

On the downside, Jony's perfectly round mouse proved to be an ergonomic nightmare. It was skittish on tabletops and difficult to orient properly; it was always pointing in the wrong direction and was too small for many adults. Users had to pinch their hands into a claw to use it, giving them cramps. Jobs was told that the mouse would be a problem, but in his race to market, he ignored the advice.

As a regular presence in the design studio, Jobs was getting excited about Columbus and the radical redesigns coming out of Jony's studio. Marjorie Andresen, the CAD expert, said the optimism and vision of

Jony's design group was infectious. "The engineers tended to be more grounded in what was possible today," she said. "The industrial designers were clearly imagining what would be possible tomorrow and in the future."

Apple set up a special factory in South Korea, where the iMac would be assembled in partnership with LG. But the manufacturing engineers raised concerns about the cost of the iMac's design. Satzger said the tooling for the case was very complex. "It was the first time we really started challenging tooling and injection molding standards."

As head of the engineering group, Rubinstein had to balance the concerns of all parties. Though Rubinstein tended to side with the engineers, Jobs usually sided with Jony and his team.

"When we took it to the engineers they came up with thirty-eight reasons they couldn't do it," Jobs recalled. "And I said, 'No, no, we're doing this.' And they said, 'Well, why?' And I said, 'Because I'm the CEO, and I think it can be done.' And so they kind of grudgingly did it."[38]

Apple's marketing chief Phil Schiller put the push-pull in context. "Before Steve came back, engineers would say 'Here are the guts'—processor, hard drive—and then it would go to the designers to put it in a box. When you do it that way, you come up with awful products."[39] But Jobs and Jony tilted the balance again toward the designers.

"Steve kept impressing on us that the design was integral to what would make us great," said Schiller. "Design once again dictated the engineering, not just vice versa."

"What Steve Jobs brought to the table was no compromise," said Don Norman. "He focused on what the product was supposed to be and he didn't listen to opposing arguments. To be sure, he listened before he made up his mind, but once he made up his mind, that was it. Before Steve, there were a lot of compromises, which led to less cohesiveness and delayed production."[40]

Some of Jobs's leadership began to rub off on Jony. "There was a kind of hierarchy of talent in the room and it was palpable. . . . [I]t was easy to see that Jony was a leader early on and that other designers looked to him for approval," recalled Andresen. She contends that while the IDg had been full of talented people in the Brunner era, she felt the atmosphere became "charged with ideas" once Jony was in charge.

Andresen remembers the time fondly. "It was fast, it was exciting and it felt like everything we were doing was really really critical. Time to market was drummed into our heads and we did anything and everything we could in order to save time getting products from concept to customer. ID—and Jonathan—was where it all started."

As the iMac neared completion, Jony's IDg worked day and night to perfect every detail. Shaking his head at the memory, Saltzer said, "We were working 24-hour shifts."

The Name Game

To help choose a name for the Columbus product, Jobs asked an old advertising buddy, Lee Clow from TBWA\Chiat\Day, to fly up from Los Angeles. Clow was accompanied by Ken Segall, one of his advertising executives. Jobs led them into a private room. In the middle of a conference table sat a big lump covered by a cloth.

Jobs whipped off the cloth to reveal a see-through plastic teardrop, the first Bondi Blue iMac. The ad men had never seen anything like it. They were stunned.

"We were pretty shocked but we couldn't be frank," Segall recalls. "We were guarded. We were being polite, but we were really thinking, 'Jesus, do they know what they are doing?' It was so radical."[41]

Jobs told them that he was betting the company on the computer, so it needed a great name. He suggested "MacMan," a name, Segall said, to "curdle your blood."[42]

The new computer was a Mac, Jobs said, so the name had to reference the Macintosh brand. The name had to make clear the machine was designed for the Internet. The name also had to be adaptable for several other upcoming products. And the name had to be found quickly, since the packaging needed to be ready in a week.

Segall came back with five names. Four were ringers, placeholders for the name he loved—iMac. "It referenced the Mac, and the 'i' meant internet," Segall says. "But it also meant individual, imaginative, and all the other things it came to stand for."

Though Jobs rejected all five names, Segall refused to give up on iMac. He went back again with three or four new names, but again pitched iMac. This time, Jobs replied: "I don't hate it this week, but I still don't like it."[43]

Segall heard nothing more about the name from Jobs personally, but friends told him that Jobs had the name silk-screened onto prototypes of the new computer, testing it out to see if he liked the look.

"He rejected it twice but then it just appeared on the machine," Segall recalled. He came to believe that Jobs changed his mind just because the lowercase "i" looked good on the product itself.[44]

"It's a cool thing," Segall remembered happily. "You don't get to name too many products, and not ones that become so successful. It's really great. I'm really delighted. It became the nomenclature for so many other products. Millions of people see that work."[45]

Over the last few years, the debate about dropping the "i" prefix has come up several times at Apple, Segall reported. "They've asked: 'Should the company drop the 'I'?' But there's a desire to keep it consistent: iMac, iPod, iPhone. It's not as clean as it should be, but it works."

Fine-tuning

Apple's factory in Singapore produced the iMac's motherboard, but the other components, including the revolutionary case, were made and assembled by LG in the specially set-up factory in South Korea.

Jony and Danny Coster visited the factory several times with engineers from the product design group to fine-tune the case molds. Visiting the factory is standard design practice, but Jony's team went above and beyond, spending far more time at the plant than is usual in order to get the iMac perfect. Most of the workers lived in dorms and ate in a big, on-site cafeteria. Jony and Danny often ate in the cafeteria, too, but stayed in a nearby hole-in-the-wall hotel. On their last visit before the iMac was released, they stayed for two weeks.

"We were in the factory most of the time, from 8 a.m. in the morning until 8 to 9 p.m. or even 1 or 2 a.m.," said Amir Homayounfar, a manager on the product design team. "They would basically bring samples throughout and the CAD guys and the tooling engineers would make further modifications and then we'd be back for more."[46]

At first, the cases came out of the molds with lots of burrs and sharp edges. "To make it basically as perfect as possible," Homayounfar said, "we would go through that iteration multiple times until Jony and Danny were satisfied with the samples. Perfection was the goal."

As launch day approached, Jony and Danny returned to the factory with twenty-eight engineers to prepare a batch of samples. The team worked long into the night, all weekend, getting the machines ready. "We were all there scrubbing these plastic cases with sandpaper to get thirty units to ship back to the U.S. for the launch of the iMac," said Homayounfar. "Thirty Apple guys plus the entire LG factory."

Jony needed thirty Apple employees at the factory because each person would check an iMac as luggage on the plane. "We flew direct from Seoul to San Francisco and an Apple truck came by at the airport and took them away. So they went from the factory to the airport, to Cupertino and to the Apple campus," said Homayounfar. "They fired them up and got Steve to come and look at them. He cherry-picked the best ones and we were ready for the launch announcement."

There were still bumps in the road ahead, however. The day before the launch, while practicing his presentation with a hastily assembled prototype, Jobs pressed a button on the front of the tray. The tray slid out.

"What the fuck is this?" he asked. He had been expecting a slot-loading drive, which were just starting to appear on high-end stereos.[47]

No one said a thing, but Jobs flew into a full-on rage. Rubinstein had chosen a tray drive to keep pace with rapidly changing CD technology. With writeable CD drives on the horizon, Rubinstein knew a slot-loading drive would put the iMac a generation behind. He insisted that Jobs had been aware of this, but Jobs became so upset, he nearly canceled the launch.

"It was my first product launch with Steve and the first time I saw his mind-set of 'If it's not right we're not launching it,'" remembered Schiller. Jobs was mollified only when Rubinstein promised to replace the CD drive in the next release.[48]

On May 6, 1998—the following day—Jobs unveiled the iMac at a packed Flint Center in Cupertino. It was the same venue where Apple introduced the original Macintosh fourteen years earlier.

The event drew a big crowd of technology press. The auditorium was buzzing with excitement as a giant inflatable beach ball bounced around and people took their seats. "I haven't seen this much energy around Apple since 1989," said Bajarin, the analyst who had been at Jobs's design meeting several months before.[49]

Jobs kicked off the show by showing a new TV ad poking fun at Intel.

On screen, a steamroller flattened several Pentium notebooks as the theme from the old *Peter Gunn* TV show blasted. The audience howled.

Jobs detailed the shortcomings of PCs for consumers: They were slow, complex and unattractive. "This is what they look like today," Jobs said, as a beige computer was projected onto the backdrop behind him. "And I'd like to take the privilege of showing you what they are going to look like from today on."[50]

He strolled over to a pedestal in the middle of the stage and pulled a black cloth off an iMac, which gleamed and sparkled under the auditorium spotlights. Jobs looked like he'd been expecting applause, but the audience, previously excited, reacted in stunned silence.

"The whole thing is translucent," Jobs gushed. "You can see into it. It's soooo cool! . . . The back of this thing looks better than the front of the other guy's."

As a cameraman walked around the iMac onstage, showing it from all angles, the audience started to respond. "It looks like it's from another planet," Jobs said proudly, eliciting a laugh from the crowd. "A good planet. A planet with better designers."

Jony and most of the design team were sitting in the audience. "I was very proud of the iMac on its release, because I had had the chance to have a lot of influence on it and had done a lot of work on many of the parts," Satzger said. "I was sitting among a crowd of Apple employees when Jobs pulled off the sheet to reveal the iMac, and all these Apple people were exhibiting surprise. I realized that none of them had seen it yet. They were working on legal, sales, operations, even software people, and none of them had been shown the machine, had been allowed to see it. That was a shock to me."[51]

The iMac began shipping on August 15, 1998. Jobs had bet the future of Apple on the machine, and, during the summer leading up to its release, Apple spent one hundred million dollars on advertising. To prime the pump with the press, Apple's PR department told reporters to expect the biggest launch in Apple history.

Apple's comprehensive advertising campaign blanketed TV, print and billboards with colorful, witty ads. The campaign emphasized the iMac's funky design and ease of use. One amusing TV ad depicted a race to set up a new iMac for surfing the Web versus a PC. The spot pitted a seven-year-old and his dog against a Stanford PhD student. (No prizes for guessing who won.) Another advertising theme, entitled the "Un-PC," contrasted the nest of cords associated with most computers with the iMac's clean, uncluttered design.

A week before its release, Apple announced 150,000 preorders for the iMac, and Apple's stock soared to more than forty dollars a share, its highest in three years. The company arranged special launch events at big-box computer stores (the first Apple stores were several years in the future), which were attended by some of the company's executives, including Jony.

Despite the hype, the first reviews were surprisingly negative, sometimes brutally so. The iMac was too radical a break with the past, they said. Tech reviewers, who, counterintuitively, tend to be rather a conservative bunch, complimented the funky design, but complained about the legacy-free technology. There was an uproar in the tech press about the lack of a floppy drive. Most articles obsessed about it. Many said the machine was stillborn without it. "The iMac will only sell to some of the true believers," wrote Hiawatha Bray in the *Boston Globe*. "The iMac doesn't include a floppy disk drive for doing file backups or sharing of data. It's an astonishing lapse from Jobs, who should have learned better . . . the iMac is clean, elegant, floppy-free—and doomed."[52]

The lack of a floppy put Jony on the defensive. "I can't give you the best Apple answer on the lack of the floppy," he said. "I can give you my answer: 'When you move on, you leave some things behind. The floppy drive, which I will argue until I'm blue in the face, is really antiquated technology. I've heard the complaints, but if there's not some sort of friction in a move forward, your step is not as consequential as you'd like to believe it is.'"[53]

Other familiar complaints about the Mac also surfaced, including the relatively high price, the incompatibility with Windows and the paucity of software compared to Microsoft's dominant platform. "Doubts about software availability make it difficult for customers to pony up the money for a Mac when a Windows-based machine is often cheaper," the Associated Press said. "And with about three percent of the computer market, the Mac is simply seen as a fringe product by many shoppers."[54]

Nonetheless, Apple's fan base was pumped. "It's just wildly different," Hal Gibson, executive director of the Berkeley Macintosh Users Group, said. "And when Apple does that, something daring, that's very exciting."[55]

Many retailers were also enthusiastic. "We'll sell lots of them. This is the sexiest computer I've ever seen," said Jim Halpin, president and CEO of CompUSA, at the time the biggest computer retailer in the United States.

The reaction from consumers was unmistakable. The iMac went on sale in August 1998 for $1,299. It sold 278,000 units in its first six weeks, and would sell 800,000 by the end of the year, making it the fastest-selling computer in Apple history. Just as Jobs had hoped and predicted, the iMac sold well to first-time computer buyers and unhappy PC users, with an impressive 32 percent of the sales going to first-timers and another 12 percent to "switchers."

Reporter Jon Fortt, writing in the *San Jose Mercury News*, noted that Apple's focus on the needs of the consumer made the iMac a hit. "What made the original iMac cool was not its color or shape. It was Apple's demonstrated willingness to open the possibilities of Internet computing to an audience that had been ignored by the brainiacs who design PCs."[56]

At the end of the quarter, Apple announced its third consecutive profit since Jobs's return; the $101 million exceeded everybody's expectations and prompted a raft of Apple-is-back stories.

A Rising Tide

The iMac saved Apple and cemented Steve Jobs's reputation as a technology seer and leading arbiter of consumer trends. Business, design, advertising, TV, movies and music would all eventually feel the effect of the iMac. The iMac was also Jony's coming-out party, the first product that gained him public attention. Overnight, Jony was celebrated as one of the world's most daring and original designers.

Over the next few years, the iMac sparked an explosion of see-through plastic products, from Swingline staplers to the George Foreman grill. It was impossible to visit stores like Target without seeing see-through cameras, hair dryers, vacuum cleaners, microwaves and TVs, aisle after aisle of translucent plastic products with bulbous, organic shapes. The trend was especially pronounced in personal electronics with portable transparent CD players, pagers and boom boxes.

"What's one of the hottest things in product design today?" asked *USA TODAY* in December 2000. "Translucence." The paper called it "electronics voyeurism."[57] Several rival companies released Windows PCs that copied the iMac. Apple sued and shut down knockoffs from eMachines and Future Power.

No longer business tools hidden away in office cubicles, computers, thanks to the iMac, became fun and fashionable. Suddenly, people were proud to put an iMac in either their living room or on the reception desk at work. According to British design historian Penny Sparke, "The iMac . . . broke the mold of computers, which hitherto were masculine objects. Then they became much more desirable objects. It was a real breakthrough."[58]

Apple's strategy of selling to individual consumers, rather than corporations, paid off. "There is a parallel here with the 1950s, when design had a lot of momentum," noted Susan Yelavich, assistant director for public programs at Cooper-Hewitt, National Design Museum, a branch of the Smithsonian Institution. "A whole new generation of products has technology as the underpinning. But, unlike in the past when technology was used in the office, today it is in and around the home." She pointed out another key distinction: The advent of the iMac meant "office products [were] being marketed to teens."[59]

The iMac shifted the computer conversation entirely. Suddenly, thanks to the iMac, prosaic details like CPU speed became less important than good looks, ease of use or custom options.

Jony argued that it was the iMac that changed the equation. "The response to the iMac makes clear that there is a widespread conception that stuff is too complicated and divorced from human concerns," said Jony. "All the attributes that are emotive have been ignored. It's about time that changed."

There were many who accused Apple of cynically designing the iMac to look different just to get attention. Bill Gates, for one, offered, "The one thing Apple's providing now is leadership in colors. It won't take long for us to catch up with that, I don't think."[60]

Jony countered that the iMac wasn't designed to look different, but the machine ended up being different as a natural consequence of the design process. "I think a lot of people see design primarily as a means to differentiate their product competitively," he said. "I really detest that. That is just a corporate agenda, not a customer or people agenda. It is important to understand that our goal wasn't just to differentiate our product, but to create products that people would love in the future. Differentiation was a consequence of our goal."[61]

Jobs had made the sudden independence of design possible at Apple.

"In most people's vocabularies, design means veneer," Jobs told *Fortune* shortly after retaking the reins at Apple. "But to me, nothing could be further from the meaning of design. Design is the fundamental soul of a man-made creation."[62]

The creation of the iMac also forged a bond between Jobs and Jony that would evolve into one of the most fruitful creative partnerships in the modern era. Between them, they changed Apple's engineering-driven culture into a design-driven one. "The real strength of the ID team became Steve's connection with Jony," said Satzger. "Without Steve being there, the places we [the ID team] were going were just crazy."

"In a company that was born to innovate, the risk is in not innovating," Jony said. "The real risk is to think it is safe to play it safe. Steve has a clear vision of what it is going to take to get back to the company's roots, what it would take to get at the essence of Apple, what it takes to structure the company to be something that can design and make new things."[63]

CHAPTER 6

A String of Hits

The historical way of developing products just
doesn't work when you're as ambitious as we are.
When the challenges are that complex, you have
to develop a product in a more collaborative,
integrated way.[1]　　　　　　　　　　—JONY IVE

Steve Jobs loved the iMac, but as soon as it hit the market, he changed his
mind about the color. In his typically binary fashion, he decided he hated
the Bondi Blue. "I love the product but we've picked the wrong color," he
told the design team. "It's not bright enough. There's not enough life to it."[2]

Jony directed Doug Satzger, the design group's official lead for color
and materials, to start a new color investigation. Satzger was given two
weeks to present new options. He found an unused room on Apple's
campus and assembled dozens of different-colored plastic items, among
them kitchen flatware, transparent thermoses, brightly-colored plastic
plates, just as the team had done for the Bondi Blue iMac. He arranged
them by color: blue items on this table, red items on another. When he
was ready, Satzger and a freelance contractor he was working with
presented the results to Jony and Steve.

It didn't go well.

"Steve walks into the room and [says] there's way too much stuff in
there," remembered Satzger. "He looks at me and he says, 'You suck.'"
Satzger could later laugh at the memory, but, at the time, it wasn't funny.
Jobs was anything but amused.

Jobs was upset because the choices were overwhelming. "We gave him way too much information, and it wasn't focused on how we could apply it to the iMac. So he looks at me and says, 'When can I see some colors on a product that's similar to the iMac?'

"I asked for three weeks. Jony looked at me and was like 'What? are you crazy?'"

The task was daunting. In the allotted time, Satzger worked like a dog to come up with nearly complete models in a variety of new colors for Jobs.

"One great difficulty that we had was transferring opaque colors into transparency," he said. "If you have a concept of a yellow, for instance, it is not very easy to make a transparent version of that yellow. What we did was take test tubes full of water and added food colors and other dyes to end up with fifteen different colors, which we took to the manufacturers to make. We did it that day, that day with Steve. And if their color house couldn't make them we went to other manufacturers."

The designers set to work building the cases and all their details, including CD drive lips, speaker enclosures, the back cover and the foot of the machine. A factory in China quickly built fifteen iMacs (with fake insides) in different colors.

Satzger chose mature, rich colors: a deep blue, "Amber beer," "Blue glue," and "green leaf." Amazingly, he met the deadline, three weeks to the day.

"Oh, my God," said Jobs when he walked into a room full of colorful new iMacs.

"Steve walked in and looked at all the models," Satzger recalled. "He took the yellow one, picked it up and placed it in the corner of the room, turned round and said, 'It looks like piss. I don't like yellow.'

"He chose some colors that he liked but then turned to us and said, 'I really like these colors, they remind me of Life Savers. But there is still something missing—a color for the girls. I want to see a pink. When can I see a pink?' So we went back again and created five different pinks in ten days, of which Steve approved one—Strawberry."

Jony was amazed at the speed with which Jobs gave his approval to the new colors. Jobs's decisions meant five different cases would have to be made at the factory, and retailers would have five new SKUs to stock. But such logistics weren't even discussed. The decision was driven purely by design: Steve Jobs wanted new colors. The logistics could be worked out later.

"In most places that decision would have taken months," Jony later remarked. At the old Apple, too, executives would have stopped to ask questions about manufacturing and distribution—not now. "Steve did it in a half hour."[3]

The new multicolored iMacs—code-named Lifesavers—were put into production and hit the market in January 1999, just four months after the original iMac went on sale. The five colors were given consumer-friendly names: strawberry, blueberry, tangerine, lime and grape. Their arrival was a significant first as the multicolored iMacs introduced the concept of fashion to an industry previously preoccupied with speeds and feeds.

The Lifesaver Macs were the first in a long series of rapid upgrades to the iMac. Over the next several years, the iMac would get faster chips, bigger hard drives, wireless networking and—perhaps most important, though few realized this at the time—an even wider range of colors and designs. With the second-generation iMac, the color options were updated to graphite, ruby, sage, snow and indigo, with some patterned machines toward the end, including "Flower Power" and "Blue Dalmatian."

Apple would continue to sell multicolored iMacs until March 2003, when they were superseded by an even crazier design, the iMac G4 (which looked like a bulbous Luxo lamp). But for four years, the iMac launch and continuing development defined the game plan that Apple would use to such devastating effect with later products like the iPod: They created a breakthrough product, then quickly and relentlessly polished it with rapid new releases. The iMac was aggressively updated, not just with new technology, but new colors and price points. In five years, with variations in chip speeds and other hardware, there were at least thirty-two models in more than a dozen colors or patterns.

The Apple New Product Process

In the months after the iMac launch, the A team also perfected a new methodology for developing products. Called the Apple new product process, or ANPP, it would emerge as one of the keys to Apple's success.

Not surprisingly, in the world according to Steve Jobs, the ANPP would rapidly evolve into a well-defined process for bringing new products to market by laying out in extreme detail every stage of product development.

Embodied in a program that runs on the company's internal network, the ANPP resembled a giant checklist. It detailed exactly what everyone was to do at every stage for every product, with instructions for every department ranging from hardware to software, and on to operations, finance, marketing, even the support teams that troubleshoot and repair the product after it goes to market. "It's everything from the supply chain to the stores," said one former executive. "It's hooked into all the suppliers and the suppliers' suppliers. Hundreds of companies. Everything from the paint and the screws to the chips."[4]

The ANPP involves every department from the outset, including functions like marketing, whose work will only be seen after the product is launched. "It's very important at Apple that the needs of the customer and needs to compete in the marketplace are considered when we create a product right from the beginning," said Apple's head of marketing Phil Schiller. "[M]arketing is an equal member of the team creating our products, along with the engineering and operations team."[5]

Modeled in part on the best practices of HP and other Silicon Valley companies, the ANPP system, which Jobs had initiated at NeXT, was perfected in the early days of Jobs's return to Apple. While such a procedure might seem to imply a hidebound approach, it was a worthy, pioneering move for Apple. One insider at the time described it this way: "It's a very well-defined process, but it's not onerous or bureaucratic. It allowed everyone to be more creative where it mattered, not less. Look at the results. Apple is a very fast company."[6]

The system applied to Jony's department, too, as the designers now had to tick off all of the steps, from investigation and concept to design and production. Sally Grisedale, former manager of Apple's advanced technology group (which worked closely with the design group), said it was the systematic documentation that set Apple's ANPP apart.

"It's all written down. It has to be. There are so many moving parts," she said. "Even when I was there, all the processes were worked out. That's why [Apple] was such a perfect company to work for, because they had booklets on how they do it, and they helped you, when building the software or the hardware. It had to be really systematic. So it was a very rude awakening for me to go a different company like Excite or Yahoo because they had none of that! Nothing written down. Like, Process? Are you kidding? Just ship it and get it out there!"[7]

Another inspiration for the ANPP was the modern engineering management system known as "concurrent engineering," which permits different departments to work in parallel (unlike the old model, under which projects get passed from one team to another in serial).

Big, complex engineering organizations like NASA and the European

Space Agency were early proponents of concurrent engineering. It's a complex but flexible methodology that tends to catch problems early, because it takes into account the entire production process and the full lifetime of the product, from manufacture to servicing and recycling. Jony has expressed admiration for the designers of space satellites, likely because of their use of concurrent engineering.

At the old Apple, the engineers would work on a product before passing it to the designers to skin it. This wouldn't work for Jobs's new Apple, with the increased primacy of the ID studio.

"The historical way of developing products just doesn't work when you're as ambitious as we are," Jony has observed. "When the challenges are that complex, you have to develop a product in a more collaborative, integrated way."[8]

Completing the Lineup

After the success of the iMac, Jobs and his A team set about filling the three empty quadrants in his 2×2 product road map. With a consumer desktop on the market, they still needed a professional desktop and portables for consumers and professionals. These machines, released over the next couple of years, allowed the company to grow and the design team to be more ambitious in pioneering new technologies, materials and manufacturing methods.

For Jony's design team, the iMac was succeeded by an assignment to create a powerful desktop computer for professional users, in particular photo editors, video editors and scientists. They were the remnants of Apple's old user base—creative professionals associated with the desktop publishing segment—who, having helped establish Apple during the late 1980s and early 1990s, stayed committed to the company and kept it afloat.

The Power Mac G3 brought the iMac design language to a tower design. Known as the "Blue and White" Power Mac, it sported a blue and white plastic case. Again, the designers incorporated a handle or, more accurately, handles, as there was one on each corner. In the case of the G3, the handles were actually designed to move the machine (rather than to make it less intimidating to users), and they were consistent with Jony's new design language.

The G3 was a quiet success. It never attracted the attention or adulation of the iMac, but it sold in respectable numbers and maintained Apple's presence in the business market, which at the time was more important than the consumer sector.

In design terms, its successor was more interesting. The Power Mac G4 tower was based around the PowerPC "G4" chip. Apple referred to the Power Mac G4 as "not only the fastest Mac ever" but also "the fastest personal computer ever."[9] The gray case also represented an interim step toward the aluminum that later came to dominate Apple's professional line.

Initially, the Power Mac G4 tower was slate gray graphite, subsequently replaced by a dazzling metallic "quicksilver" case. Satzger remembered working on the quicksilver case, which taught him an important lesson about the new Apple.

At the last minute before the machine went to production, there had been a change in various hardware features, as well as in the color of the case. In the rush to build it, there was a color mismatch between the doors on the front face and the rest of the case. Jobs, of course, didn't like it. Satzger pushed back, saying that there wasn't time to fix it. According to Satzger, "Steve said simply, 'Don't you think you owe it to yourself and to me, to do better?' I said yes and we went back and did it again, and it was better. It was always better."

The quicksilver case was also the cause of a big fight between Jony and Rubinstein. Jony wanted special screws on the handles that had a particular shape and finish. But Rubinstein said the cost would be "astronomical" and would delay the machine's delivery. Delivering products on time was Rubinstein's responsibility, and so he vetoed Jony's screws. But Jony went over Rubinstein's head to Jobs and around him to the engineers in product design.

"Ruby [Rubinstein] would say, 'You can't do this, it will delay,' and I would say, 'I think we can.' And I would know, because I had worked behind his back with the product teams," Jony later recalled.[10]

The argument over screws illustrated a growing rift between Jony and Rubinstein. Over the next several years, their clashes would become more frequent and more fraught.

"[Jony's] focus was design," said Rubinstein, speaking by phone in 2012. "Design was all he cared about. And so although design is extremely important, we also had to tackle electrical engineering, manufacturing, service and support. There are many different constituencies that all had a voice—not necessarily a vote, but a voice—and my job was to balance everybody's needs. And at some point, compromises always had to be made."[11]

Not in this case, though. Jony got his way: The screws on the G4 were made from highly polished stainless steel.

Reinventing the Laptop

The design team next turned their attention to the third quadrant of Steve's 2×2 product plan to be filled: a portable for consumers.

"The brief was simple," said one of the designers. "Bring the iMac to laptops."

The initial ideas for the iBook were all over the map. They came from brainstorming sessions, not from focus groups or market surveys. "We don't do focus groups—that is the job of the designer," said Jony. "It's unfair to ask people who don't have a sense of the opportunities of tomorrow from the context of today to design."[12]

At the time, laptops tended to be boxy, black and starkly utilitarian. "We were given a lot of freedom," Satzger said, though everyone understood the new machine would echo the iMac's curves and colorful, translucent plastic body. What emerged looked nothing like its competitors, not only in its use of curvaceous lines but because it screamed color and fun. Jony initially drew its curvy "clamshell" design that resembled an undersea creature, then handed it off to Chris Stringer to be the design lead.

The designers also incorporated a clever innovation, novel for computers at the time: The iBook "awakened" as the lid was lifted. This feature required the design team to labor on a latchless mechanism to keep the lid tightly closed when the iBook wasn't in use, because the last thing they wanted was for the computer to wake up in someone's knapsack and drain the batteries.

The shell also had an integrated carrying handle, which made the device look like a colorful plastic purse. A handle on a laptop made sense, just as it had on the Macintosh SketchPad that Jony helped design for Bob Brunner's Juggernaut project. But again, it served a dual purpose: adding portability while encouraging a connection with the machine, making it less intimidating.

"The iBook has been designed to encourage users to touch it," Jony explained. "The use of curved surfaces and rubberized materials give it an intimate, tactile feel."[13]

The manufacturing of the iBook—case, handle and lid—all introduced serious challenges. To start, the iBook's case was made from a hard polycarbonate plastic bonded to thermoplastic polyurethane, the latter a rubbery compound that softened the case's edges and made it resistant to bumps. The polycarbonate case was also bonded to the guts inside. "It was a question of shape and layering—layering plastics and sheet metal below," said Satzger. "But the product was massive. There were lots of manufacturing challenges. We spent a lot of time in Taiwan working on layering materials. There was a ton of problems."

The complex clamshell shape created a huge molding headache. To make the shell and get it out of the mold afterward, the mold had to pull apart in multiple directions and, initially, the case was plagued with micro-cracks that formed as the plastic cooled. Nor was the carrying handle easy to manufacture. It was made from a special plastic from DuPont called Surlyn, molded over a magnesium core to give it strength. Surlyn is a tough, impact-resistant plastic used in golf balls, but to make a handle of it required a molding technique called metal injection molding. First, the metal part was inserted into the molding tool; then the colored Surlyn plastic was molded around it. But metal and plastic have different cooling rates, so the part would shatter as soon as the mold was opened. The team spent weeks at the factories in Asia tweaking the molds and chemical mixes of the different plastics, but eventually overcame all of the problems.

The game-changing latchless lid also proved to be an issue—ultimately solvable—which Jony's team labored for months to figure out. They came up with the solution of making a special hinge that held the lid tightly shut when closed. Getting rid of the latch was not initially seen as a "wow" factor but part of the team's relentless drive to make products with fewer and fewer parts—again, a defining characteristic of Jony's design vision. "Less parts means better tolerancing and better part-to-part relationships," said one designer. In other words, the product goes together better.

These features—the rubbery skin, the carrying handle and hinge—and the consequent complexities delayed the iBook for months, but advance word on the iBook was strong. The new machine was much sought after, and electronics stores started taking preorders in advance of its debut.

When the iBook did launch, one wag said it looked like "Barbie's toilet seat," a name that stuck.[14] But the iBook was quickly a big hit with consumers, students and educators alike. More than a quarter million units shipped in the first three months after its release. Over the next few years, different enhancements of the clamshell iBook were issued, adding more color choices, extra memory and FireWire ports.

The iBook would also gain a place in history for popularizing Wi-Fi, the now ubiquitous wireless networking technology. Apple didn't invent Wi-Fi, but it was the first computer maker to recognize its potential, like it did with USB ports on the iMac. While Wi-Fi was available for other laptops, it required an add-on card with an ugly antenna sticking out the side. The iBook neatly solved this problem by providing built-in Wi-Fi.

While the iBook was in development, Apple started looking at home networking. The Internet was rapidly gaining users, and it was obvious that consumers would need networking technology at home. Apple's competitors were looking to offer answers too: Compaq was pushing networking through home power outlets and Intel was considering phone jacks. "We looked at it and thought, 'These are really stupid ideas,'" said Phil Schiller. Schools were a critical Apple market, and neither power-line networking nor individual phone lines would work there, so the company took another approach.

Apple had engineers involved with standards bodies (the committees that standardize technologies, like Bluetooth or USB, across the industry), and one of them alerted the executives to a new wireless networking technology called 802.11. As Schiller remembered, "We decided, really fast track . . . to change the physical design of all of our products to include antennas and card slots and to make a complete holistic solution to make 802.11 come out." They decided to call their system of networking cards and base stations "Airport."[15]

The clamshell model of the iBook was discontinued in 2001, replaced by a reinvented version of the notebook in white polycarbonate. But the iBook changed the game. Many of its innovations live on in products today, including the placement of the interface ports on the sides instead of the back; the latchless lid; and, of course, Wi-Fi, now standard on every laptop, tablet and smartphone.

With the success of the iMac and iBook, power at Apple inevitably shifted toward Jony's industrial design group. Rubinstein was forced to bring in a stream of new engineers who were able—and willing—to execute the designs issuing from Jony's group.

"There was a lot of turnover," said a former hardware executive. "We more or less replaced the entire mechanical engineering group. A lot of old-timers quit. They couldn't take the pace. We reduced product development from three years to nine months, made it one of the fastest companies in tech."

Rubinstein elaborated: "We brought in new people to run mechanical engineering—the product design group, PD—so that ID would have partners who could execute the designs that they were coming up with. And likewise we also brought in Asian suppliers who could execute the designs, since it is equally essential to have the requisite manufacturing capability."[16]

In the old Apple, the engineers called the shots. In the new Apple, the product design team, which was responsible for making sure designs can actually be manufactured, deferred to the design group.

"ID has the final say on everything," said Amir Homayounfar, who worked at Apple for ten years, ending his career as a program manager in the product design group. "We were working for them."

Marjorie Andresen went further. According to her, Jony's group was becoming the most powerful voice in the company.

"The biggest thing you had to understand about working with ID was that it was never an option to tell them no," she said. "Even if what they wanted to do seemed expensive, ridiculous or even impossible you had to make it happen. . . . Whatever it took to get the job done."

Completing the Matrix

After the iMac, iBook and Power Mac, the last machine in Jobs's 2×2 product plan was a notebook for professionals.

Jony instructed De Iuliis and two other designers to rethink the professional laptop just as they had the consumer laptop. He wanted a new user experience, an honest use of materials and a machine that was easy to manufacture. He set them up in a special studio off site in a warehouse in San Francisco where they installed thousands of dollars of computers. Just six weeks later, they presented Jony with a notebook that fulfilled two out of three of his requests.

The Titanium PowerBook G4 was the lightest and thinnest full-featured notebook yet to appear on the market. It had the first wide-screen laptop display, a big, beautiful expanse of screen that was ideally suited for working on high-end professional software, which often features numerous floating palette windows. But it would be a challenge to manufacture, and the new PowerBook would have an enclosure made from formed titanium shells separated by a plastic gasket. The whole machine was made rigid by a complex internal frame combined with several strengthening plates.

After coming up with the basic concept in just six weeks, Jony's team spent months sweating every detail. The PowerBook featured a clever latch for the screen that descended from inside the lid as it was closed. To the delight of users, the latch seemed to pop out as if by magic, appearing at just the right time as the lid was nearly shut. New owners would open and close the lid obsessively just to see the latch appear and disappear.

The latch mechanism used a small magnet in the bottom half of the PowerBook's body that pulled the latch out of a thin slot in the lid. It was a harbinger of things to come, as magnets would be used in a lot of clever ways in subsequent products, including the iPad 2, which would be awakened and put to sleep by a magnetic "Smart Cover." One of the later flat-screen aluminum iMacs even had its screen attached by magnets, which allowed easy access to its guts.

Like the power button on the back of the original Macintosh, the PowerBook's magnetic latch was the kind of detail that turned a good product into a great one. As Jerry Manock, the designer of the original Macintosh said, it's the artisanal details that count. Jony would agree.

"The decisive factor is fanatical care beyond the obvious stuff; the obsessive attention to details that are often overlooked," said Jony. [17]

Years later, designer Chris Stringer described the ID studio's obsessive attention to small things. "We're a pretty maniacal group of people," he said. "If we design a button, there might be fifty models of the home button or a volume switch. We look at the edge detail and [ask] how far out does it protrude? Does it have a shaft? Is it round? Is it metal? Is it plastic? The size, length, width, height. Every single detail is very cleverly crafted." [18]

Detailing was a continual part of the design process, not something done at the end to make a product look pretty. It was as integral as anything else, and typical of Jony's approach. "It doesn't go from thought to sketch to model to production even though, in simplistic terms, that is the general sequence of events," said Stringer. "We'll go back and forth. We'll go all the way to model, we'll go to working with the PD and operations groups on the engineering side."[19]

In design, details like buttons and latches that make a design pop have a name: They're called "jewelry." In the auto industry, door handles and radiator grilles have the same name and the same effect. The new Apple products took these elements to a new level. "We really focused on the jewel pieces," said Satzger. "We strove for really high quality. We wanted beautiful finishing, really high-quality surfaces on them."

The magnetic latch on the Titanium PowerBook was a good example of jewelry. To release the latch, there was a high-quality stainless steel button that when depressed would pop open the lid a bit, allowing users to get their fingers underneath it to open it. It was another artisanal touch that delighted owners, much the same as the popping lid Jony designed for the Lindy Newton.

Jony's team instructed the supplier of the PowerBook's power button to make several samples before getting the contract. Each sample contained twelve slightly different power buttons, all machined in stainless steel. "You can barely see the difference between them," said Satzger, smiling apologetically at the group's fastidiousness. "The subtleties are crazy."

Jony predicted the PowerBook would be a hit when it was announced at the 2001 Macworld Expo in San Francisco. "People will have a visceral reaction to its weight and volume," he said.[20] Indeed they did. It sold out immediately and remained a hard find for months. The factories couldn't keep up with demand.

Although the PowerBook was an expensive machine, it attracted many new customers to the Apple platform, including geek cognoscenti. It showed up at a lot of high-tech conferences, and became the preferred machine of alpha geeks like Linux creator Linus Torvalds. It also did a lot for rehabilitating Apple's reputation among movers and shakers in the tech industry. The iMac had been great, but it was a cute plastic toy for ordinary consumers. The Titanium PowerBook, on the other hand, was regarded as a serious machine for professionals.

It was also the design team's first foray into metal and advanced metallurgical manufacturing techniques. Titanium is notoriously difficult to work with. Bare, untreated titanium has a beautiful luster, but it picks up fingerprints and scratches easily. Jony chose to paint the PowerBook, but over time, the paint wore off around the keyboard and palm rests, which attracted complaints.

Despite its popularity, there were other problems too. The case's complex internal frame was made out of several different metals, and to make things like the magnetic latches work, parts of the case had to be made out of steel. The more parts, the more materials, the more problems. Over time, the different parts of the case tended to separate as the machine took knocks and bumps. Eventually these problems would lead the design team to pioneer a radical new manufacturing technique for its portables.

Along with the exotic metal PowerBook, Jony's design team created a new plastic portable for the consumer and education markets. The new version of the iBook was designed in bright white plastic and had two USB ports, attributes memorialized in its name: Dual USB "Ice" iBook.

For the Ice iBook, Jony's continued concern for durability prompted him to combine a polycarbonate shell with an internal magnesium frame. Critical components such as the hard drive were shock-mounted with rubber gaskets, like the engine in a car. Elements susceptible to damage, such as doors, external buttons and latches, were eliminated. Instead, the entire product was almost hermetically sealed by its external surfaces. Even the sleep-state light-emitting diode (LED) indicator did not penetrate the external skin and only became visible when the unit was asleep, as it gently cycled from dim to bright.

Thanks to a clever L-shaped hinge, the display opened to rest unexpectedly far back from the keyboard, making the product feel expansive and accessible when in use but compact when closed. "When the iBook's closed, it's this smooth, rugged little pod," said Jony in a promo video. "But as you open it, the geometry of the hinge moves the display away and down, which reveals this full-sized keyboard and large comfortable palm rest."[21]

The Ice iBook was made of transparent polycarbonate, with a white coating of paint applied to the inside surface. The transparent outer shell created a "halo" around the product, which gave the surface a surprising depth. It also made the product appear smaller than it actually was. It was scratch resistant, because the paint was on the inside. Painting the inside surface of the plastic may have been inspired by Jony's experiment at RWG, where he painted slides with gouache to create his spectacular mock-up sketches. The halo would be a popular effect used on many other products, most notably the iPod, and it persists in the glass screen of the latest iPhone and iPad.

A clean, plain, small rectangular box, the Ice iBook cemented the shift in Jony's design language from multicolored plastics toward plain black-and-white polycarbonate designs. (Although the first white computer was technically the "Snow" iMac in the summer of 2001, the shift to white plastic really gained notice with the Ice iBook, which also looked utterly unlike any other laptop at the time.)

"The new iBook is clearly from the same family as the PowerBook G4, but it certainly has its own distinct character," said Jony. "It's warmer. It's happier. I really think it's a much friendlier design."[22]

Jony would transition most of Apple's consumer products, including the iMac and iPod, to black-and-white polycarbonate casings. Most of the professional products, on the other hand, were redesigned using anodized aluminum.

Power Mac Cube

In 2000, having filled all four of the 2×2 quadrants, Jony and the design group attempted their most ambitious product to date: the Power Mac Cube.

The Cube was the team's first shot at the ultimate computer, an attempt to cram the power of a desktop computer into a much smaller case. Jony saw putting a lot of components into a tower design as lazy: Why give the consumer a big ugly tower just because it's the easiest option for the engineers and designers? They aimed to make the new machine by combining untested plastic casting with advanced miniaturization. Like many of Apple's products, it was an exercise in simplification, removing everything that could be removed. It represented a major breakthrough in miniaturization, innovative design thinking and manufacturing.

The Cube was actually a rectangle formed from a single piece of crystal-clear plastic that was translucent at the base, giving the impression that the eight-inch Cube was suspended in air. It had a vertical slot-load DVD drive on the top, which popped up the DVD like a piece of toast. Some compared the Cube to a box of Kleenex. The analogy greatly amused Jony and the designers, and they took to using empty Cubes in the design studio as tissue dispensers.

The Cube used air convection for cooling instead of a noisy fan. Air entered through vents in the bottom and cooled the chips inside, exiting through vents in the top. It operated in virtual silence.

Like the Power Mac G4 tower before it, access was a key consideration. The guts of the G4 Cube were designed to be easily removed for access to internal components; its entire core could be pulled out through the bottom with a beautifully made pop-up handle. To turn the Cube on, there was a touch-sensitive button that appeared to be printed on the surface of the transparent case. It seemed magical, as though the button floated in air, detached from the computer with no visible means of operation. It was an early use of capacitive touch (the technology that would eventually make the iPhone possible). Customers loved it.

The new machine would be configured with a 450 MHz G4 chip, 64 MB memory and 20 GB storage. Priced at $1,799, the basic model included an optical mouse, pro keyboard and Apple-designed Harman Kardon stereo speakers, but no monitor. Also available exclusively on the Apple online store was a higher-spec model G4 Cube with a more powerful processor and additional memory and storage, which went for $2,299.

"The Power Mac G4 Cube was a breakthrough product," said Satzger. "It contained a lot of interesting new technology and beautiful mechanics. It was really exciting."

Some customers went crazy for the Cube. The Cube looked "sophisticated and expensive," said the Ars Technica Web site.[23] "Holy s**t, they've done it again," said Lee Clow, chief creative officer of TBWA/Chiat/Day.[24] But the public reaction to the new machines was cooler than Jony and Jobs had hoped.

Consumers viewed the Cube as basically a mid-range Power Mac G4 tower at a higher price. It was $200 more expensive than a comparable G4, and didn't come with a monitor. The price was a lot higher than anything in the Windows world.

It also suffered from its own *Doonesbury* moment: The transparent case developed hairline cracks, an issue that got a lot of attention in the press. On some machines, tiny cracks appeared in the clear plastic case, especially around the DVD slot and a pair of screw holes in the top. It was a relatively minor cosmetic flaw, but it drove some customers crazy. "They are the worst kind of cosmetic problem," wrote the Ars Technica Web site in its review. "Something that is not 'important' enough to really fix, but which will grate on those that care deeply about the appearance of their hardware . . . the very same people that are most attracted to a system like the Cube!"[25]

In September 2000, just a few months after the Cube's debut, Apple announced sales were slower than expected. It was later revealed that Apple sold a paltry 150,000 units, only one-third of the volume Apple had projected. At the end of 2000, Apple reported earnings that were "substantially below expectations" for the last quarter, with a $600 million revenue shortfall.[26] It was Apple's first unprofitable quarter in three years.

For Apple watchers, the news was ominous. Despite a string of hits, Apple was still on shaky ground, battling powerful foes like Microsoft and Dell, which were at the peak of their power. "Frankly, I can't say I'm surprised by these numbers at all, but they do show a much bleaker picture than what has been depicted in the press," said Kevin Knox, an industry analyst at the Gartner group. He went further, concluding "They are disastrous."[27]

In July 2001, Apple issued a press release saying the Cube had been "put on ice." It wouldn't continue to be sold, but it wasn't officially discontinued either; it was suspended. The release said there was "a small chance" an upgraded model of the computer would be introduced in the future. That never happened and, five years later, the Cube was replaced by the Mac mini, a much cheaper "headless" Mac, which clearly targeted first-time, budget-conscious consumers.

For Jony's group, the Cube didn't represent all bad news. Though the machine performed poorly in the marketplace, it had its admirers internally because it represented breakthroughs in manufacturing techniques and miniaturization.

The machine reflected a new mastery of packing desktop components into laptop-sized spaces, which would be crucial in making the dome-shaped iMac and later flat-screen models. Just as important, it pushed Apple into new manufacturing techniques that would benefit later products like the iPod. Satzger explained: "Basically, we did not accept standard molding practice in plastics. We started getting into machining plastic. On the Cube, the screw holes and the vent hole were precision machined."

Such machining would come to define products like the MacBook and iPad, which are machined out of slabs of aluminum. The Cube was an early foray into machining products on a mass scale. In a wider context, the machining experiments represented a fundamental shift in how products are mass produced.

"For a long time at Apple, the engineering team often told the designers, 'You can't do that,'" said Satzger. "But the design team challenged everything—with plastics, metal, every material."

Even though the Cube didn't sell well and even became a symbol of form over function, its creation spoke for the growing power and influence within Apple of Jony Ive and his design team.

CHAPTER 7

The Design Studio Behind the Iron Curtain

> When Steve came in, he wanted the conversation to be between him and the person he was talking to. . . . [W]e would turn the music up so that his voice would carry directly to one person only. You really couldn't hear what he was saying.
>
> —DOUG SATZGER

On February 9, 2001, after the hubbub of Macworld Expo had died down, the industrial design studio moved from the building on Valley Green Drive (across the road from Apple's main campus) to a large space inside Apple's HQ, where it remains today. The new design studio was put on the ground floor of Infinite Loop 2, known internally at Apple as IL2.

The move was, in part, symbolic. Bob Brunner had set up the original studio across the street to give it some independence from the rest of the company. Now ID would inhabit space at the heart of Apple, allowing Jobs to work more closely with Jony and his team. It affirmed for all to see the elevated status of design within the company. In Jony's words, ID was now truly "close to the heartbeat of the company."[1]

The move was a big one logistically, too, since the studio had become home to big machinery and prototyping equipment. The new quarters were carefully crafted for Jony's designers, with every piece of furniture—the tables, the chairs and even the glass—made to order.

A large space, the studio occupies most of the ground floor of IL2. Security is extremely tight, and a wall of large frosted windows across the bottom of the building prevents anyone from getting a peek inside.

Inside, the studio is divided into several different spaces. To the left of the entrance is a well-equipped kitchen with a large table where Jony's team conducts their biweekly brainstorming sessions. To the right of the studio's front entrance is a small, rarely used conference room.

Opposite the front entrance is Jony's office. A glass cube measuring about twelve feet by twelve feet, it's the only private office in the studio. The front wall and door are made of glass with stainless steel fittings, just like the ones in Apple's stores. Except for a small shelf system, Jony's office is bare with plain white walls, featuring no pictures of his family or design awards, just a desk, chair and lamp.

His leather chair is a Supporto chair from the UK office furniture manufacturer Hille. Designed in 1979 by the award-winning designer Fred Scott, the leather and aluminum chair is recognized as a design masterpiece. Jony himself cited it as one of his favorite designs ("the Supporto is a wonderful chair," Jony told *ICON* magazine),[2] and he selected it for the new Industrial Design Centre in Cupertino, California, and for his designers, all of whom sit at Supporto desks with leather chairs.

Jony's desk was custom-made by London-based designer Marc Newson, one of his best friends. The desk is usually bare except for his seventeen-inch MacBook and several colored pencils used for drawing, which are typically arranged neatly on his desk. He doesn't use an external monitor or other peripheral equipment.

Directly outside Jony's office are four large wooden project tables that are used to present prototype products to executives. This was where Steve Jobs gravitated when he visited the studio. In fact, the studio setup gave Jobs the idea for the big open tables in the Apple stores. Each table is dedicated to a different project—one for MacBooks, another for the iPad, the iPhone and so on. They are used to display models and prototypes of whatever Jony has to show Jobs and other executives. The models are covered at all times with black cloth.

Next to Jony's office and the presentation tables is a large CAD room, also fronted by a glass wall. The CAD room is home to about fifteen CAD operators ("surface guys"). If the designers want to see what a CAD model looks like as a real object, they'll send the file to a CNC milling machine in the machine shop next door.

The machine room, aka "the shop," is at the far end of the space. The shop is divided internally into three rooms by more glass walls. At the front are three big CNC machines. These big hulking milling machines are capable of crafting anything from metal to RenShape. They have covers that contain scrap materials, making them "clean" machines. Behind them are the "dirty" machines, namely, various cutting and drilling machines that can create a mess. The so-called dirty shop is sealed behind glass. Next to it, on the right, is a finishing room with fine-sanding machines and a big paint-spraying booth about the size of a car.

The shop is used by the design team to make models of upcoming products to quickly validate ideas. "They'd get a CAD file for surfacing; they would create a tool path based on the surfaces, do all the setups and run a part," said Satzger. As well as modeling the basic shape of a product, the team will make models of details: the corner of a product or its buttons, for example. Often, Jony's team makes hundreds of models, just like Jony did in college. As the product development process progresses, the design team will outsource model making to an outside specialist firm.

Jony's office, the presentation tables, the CAD room and shop are all to the right of the front entrance. On the other side, an opening near

Jony's office leads to the area where the designers work. In the large open space, lined by the long wall of frosted windows, the designers work at five large tables, subdivided by low dividers. The space is messy and chaotic, with boxes, parts, samples, bikes and toys everywhere. The atmosphere feels light, fun. "Someone might be skateboarding in there, doing jumps, or Bart Andre and Chris Stringer kicking a soccer ball," said Satzger.

Music is an important part of the design studio's atmosphere. There are about twenty white speakers in the room, with a pair of thirty-six-inch-high concert subwoofers. "When you walked into this concrete and steel, highly reflective room, the sound was immediately deep and loud," Satzger remembered. "All kinds of music from around the world is played. It's really lively. We had so much music on that thing, you could pick anything."

Jony is a big fan of techno—music that drove Jony's boss, Jon Rubinstein, to distraction. "They played loud techno-pop in the design studio," he said. "I like quiet so that I can focus and think properly. But the ID guys liked it."

"The energy of that room, the noise of the room made me work a lot better," said Satzger. "I hated sitting back in my little space. . . . [F]or me, the louder the noise the better."

Steve Jobs liked the music too, and often turned up the music when he visited, but for more than its sonic pleasures. "When Steve came in, he wanted the conversation to be between him and the person he was talking to," said Satzger. "In all these open spaces, if it's quiet, it's really easy to hear what people are talking about. When he came in, we would turn the music up so that his voice would carry directly to one person only. You really couldn't hear what he was saying."

The studio seemed to have a visible effect on the intense Jobs. "Steve in the ID space was a different person. He was a lot more relaxed and interactive," Satzger said. "Steve had moods all the time. There were always things that were changing in how he approached people. But when he came in the ID space, he was always really comfortable."

Jobs spent a lot of time in the studio, but when he was away, Jony used his absence as an opportunity to get work done for the absent master. "When he went away, we would do 150 to 200 percent more work," Satzger explained. "It was a good chance to blast stuff out and put new work, new ideas in front of him when he came back."

The Iron Curtain

Jobs significantly beefed up security when the design studio moved onto the main campus. Because the studio is Apple's ideas factory, Jobs wanted no leaks. He knew security was sometimes lax on Valley Green, and that visitors were occasionally buzzed in by whoever was around. Jobs was determined that wouldn't be the case in the new location.

The vast majority of Apple's employees are barred from the company's design lab. Even some members of the executive team are forbidden from entering the studio. Scott Forstall, for example, who rose to be head of iOS software, wasn't allowed to visit. His badge wouldn't even open the door.

Few outsiders have been inside the studio. Jobs would occasionally bring in his wife. Walter Isaacson was given a tour, but he described only the presentation tables in his biography of Jobs. The only known photograph of the studio was published in *Time* magazine in October 2005.[3] The photo shows Jobs, Jony and three executives at the studio's wooden project tables, the shop in the background.

There's also some subtle subterfuge in the media with the ID studio. Jony occasionally gives interviews on Apple's campus in an engineering workshop full of CNC milling machines. It's been identified as the design studio, but it's actually an engineering workshop nearby.

The secrecy extends well beyond physically protecting the products within Apple's own studios. When working on new products, the software engineers have no idea what the hardware looks like and, conversely, the engineers have no idea how the software works. When Jony's team was making prototypes for the iPhone, the designers worked with a picture of the home screen with dummy icons.

Although all departments have their proprietary information, the secrecy is evidently tilted more in one direction. Apple's most secretive department is Jony's group. "It's locked down," said Satzger. "People know not to talk about their work and what was going on inside Apple to the wrong people." Who are the "wrong people"? Essentially anyone but direct colleagues. Even Jony is forbidden from telling his wife what he's working on.

A former Apple engineer who worked closely with Jony's group in the product design team said the secrecy could get exhausting. "Out of everything I've ever done in my life, I've never seen a more secret environment than working there," he said. "We were constantly under threat of losing our jobs for revealing any shred of anything. And even within Apple, your neighbors often didn't know what you were working on. . . . The secrecy was like a gun to your head. Make one false move and we'll pull this trigger."[4]

Apple's obsessive secrecy has also meant that the designers have gotten almost no press and very little public recognition. They've won almost every award and are certainly admired in design circles, but to the general public, they remain largely anonymous. There's surprisingly little resentment about the lack of credit. The team is used to it and Jony is gracious in sharing the recognition their products do receive. Though he typically gets the awards, he always talks about the team. As one observer wryly noted, the only time Jony says "I" is when he's talking about the iPhone or iPad. The move is as much to shield the individuals on the team, however, as it is to give due praise. According to Rubinstein, "Although in the press Jony gets all the credit, in reality his whole team does a lot of work. . . . They all contribute tremendously with great ideas."[5]

The design team doesn't take offense at the lack of individual public praise. "We took it as, we're all getting credit," said Satzger. "[Apple] always says, 'the Apple design team,' but Steve never wanted us to be in front of the camera. They blocked headhunters and search people. Because we were blocked from facing media and hidden from headhunters and so on, we called ourselves the 'ID Team that's Behind the Iron Curtain.'"[6]

As the principal inventors at Apple, Jony and his group conceive and create new products, refine existing ones and do some fundamental R&D, though they are not the only R&D group in the company (there is no distinct R&D department). The sixteen-odd designers focus on refining and improving Apple's products and manufacturing processes. By comparison, Samsung has a thousand designers working in thirty-four research centers around the world. Of course, Samsung makes many more products than Apple (including some components for the iPhone and iPad).

Stringer describes the role of an industrial designer at Apple as "to imagine objects that don't exist and to guide the process that brings them to life. And so that includes defining the experience that a customer has when they touch and feel our products. It's managing the overall form and the materials, the textures, the colors. And it's also working with engineering groups to, as I say, bring it to life, to bring it to the market and to building the craftsmanship that it absolutely needs to have to have that Apple quality."[7]

Jony's ID group has become a tightly knit team, as many of them have worked together for decades. They no longer design Apple's products alone, but each product has a designated design lead, the designer who does most of the actual work, plus one or two deputies.

Weekly meetings ensure the design process is collaborative. Two or three times a week, Jony's entire team gathers around the kitchen table for brainstorming sessions. All of the designers must be present. No exceptions. The sessions typically last for three hours, starting at nine or ten a.m.

The brainstorms begin with coffee. A couple of the designers play barista, making coffee for the group from a high-end espresso maker in the kitchen. Daniele De Iuliis, the Italian from the United Kingdom, is regarded as the coffee guru. "Danny D was the person who educated us all on coffee and grind and the color of the crema, how to properly do the milk, how temperature is important and all that stuff," said Satzger, who was one of his keenest disciples.

When it's time to get down to it, the brainstorms are freewheeling, creative roundtables where everyone is expected to contribute. Jony runs the brainstorms, but he doesn't dominate them.

"Jony's always been involved in every design session," said one designer.

The brainstorms are very focused. Sometimes it's a model presentation, sometimes the detail of a button or speaker grille is discussed, or the group hashes out whatever design problem Jony's group is working on.

"[We] discuss our objectives, and so we can just be talking about what we would want a product to be," said Stringer. "That ordinarily becomes sketching, so we'll sit there with our sketch books and sketch and trade ideas and go back and forth. That's where the very hard, brutal, honest criticism comes in and we thrash through ideas until we really feel like we're getting something that's worth modeling."[8]

Sketching is fundamental to their workflow. "I end up sketching everywhere," said Stringer. "I'll sketch on loose-leaf paper. I'll sketch on models. I'll sketch on anything I can put my hands on, quite often on top of CAD outputs for want of better things to do." Stringer likes CAD printouts, he's said, because they already have the shape of the product. "You're working with something that already has the perspective set up and the views in a way that you can sort of add in lavish detail upon them," he said.[9]

Jony is also an inveterate sketcher. He is a good at it, but emphasizes speed over detail. "He always wanted to get a thought down on paper so that people could understand it really quickly," said Satzger. "Jony's drawings were really sketchy, with a shaky hand. His drawing style was really interesting."

Jony's sketchbooks are "really cool," remembered Satzger, but he regards the artists of the group as Richard Howarth, Matt Rohrbach and Chris Stringer. "Richard Howarth would come in saying he had a crap idea and 'you guys are going to hate it' but then shared these amazing sketches."

When the group was designing the iMac, the table was covered with loose sheets of copy paper for sketching on, but Jony's group moved on to use sketchbooks, often hardbound volumes from Cachet by Daler-Rowney, a small British company. The studio's office supply stockroom is stacked with them. With bindings made of high-quality canvas, they don't fall to pieces. Howarth and Jony chose blue sketchbooks about three times as thick as the Cachet sketchbooks, with ribbon markers.

The sketchbooks make it easy to go back and look at earlier ideas, which is vital, as the group's practice is to document everything generated during a brainstorm. These sketchbooks would later became a contentious issue at the *Apple v. Samsung* trial.

A lot of sketching happens in these weekly sessions. At the end of the brainstorm, Jony will sometimes instruct everyone around the table to copy their sketchbooks and give the pages to the lead designer on the project under discussion; Jony will later sit with the lead designer and carefully go through all the pages. The lead and his two deputies will also pore through the pages trying to find ways to integrate new ideas.

"Some days I'd be engaged and have ten pages of stuff," Satzger remembered. "Sometimes you could feel when a designer wasn't engaged in the material, when they weren't filling up pages of things."

Model Making

Before presenting ideas to Jobs or another executive, the making of realistic mock-ups are outsourced to a model shop. The goal is to create models that look as much as possible like a final, finished product, which requires specialist equipment and skills. Jony's group frequently uses Fancy Models Corporation, a model-making company based in Fremont and run by Ching Yu, a model maker from Hong Kong. Most of the iPhone and iPad prototypes were made by Fancy Models. Each model costs in the range of $10,000 to $20,000. "Apple spent millions on models made by that company," said a former designer.

By contrast, the CNC machines at Apple, though capable of making pretty refined models, are used mainly for fairly crude early models, or parts that are needed quickly, such as plastic shapes and smaller aluminum bits. The Apple CNC machines are rarely used to produce final models.

The models play a crucial role in deciding on the final design. When designing the Mac mini, Apple's relatively inexpensive "headless" Mac,

Jony had about a dozen models made up of different sizes. The models ranged from very large to very small. He lined them up on one of the presentation tables in the studio. "We were there with some of the VPs and Jony," said Gautam Baksi, a former Apple product design engineer. "They pointed at the smallest one and they said, 'Well, obviously that is too small; that kinda looks ridiculous.' Then they pointed at the other side and said, 'Well, that's too large; no one wants a computer that big. How do you find what's right in the middle?' And they talked through that process."

The decision about the size of the case might seem trivial, but it would influence what kind of hard drive the Mini could contain. If the case were large enough, the computer could be given a 3.5-inch drive, commonly used in desktop machines and relatively inexpensive. If Jony chose a small case, it would have to use a much more expensive 2.5-inch laptop drive.

Jony and the VPs selected an enclosure that was just 2 mm too small to use a less expensive 3.5-inch drive. "They picked it based on what it looks like, not on the hard drive, which will save money," Baksi said. He said Jony didn't even bring up the issue of the hard drive; it wouldn't have made a difference. "Even if we provided that feedback, it's rare they would change the original intent," he said. "They went with a purely aesthetic form of what it should look like and how big it should be."[10]

Jony's Role

Jony's role at Apple evolved, as he became more managerial than design driven. Jony both ran the group and recruited new members. He was the conduit of information between the design group and the rest of the company, especially at the executive level. He worked very closely with Steve Jobs when he was alive—and now with Apple's executives—to select what products to work on and what directions they should take. Nothing is done without his input, whether it's the color of a product or the detail of a button. "Everything is reviewed by Jony," said one of the designers.

According to Rubinstein, "Jony is a good leader. He is a brilliant designer, and his team respects him. Jony has very good product sense." Satzger shared this assessment.

"Jony is very effective as a leader," Satzger reported. "He is a soft-spoken English gentlemen who had Jobs's ear." It was clear in the past that, in many ways, Jony was the hand implementing Jobs's vision. If Jobs didn't like something, he'd say so, but that was the only direction he gave. His feedback was always non-directive. He never suggested how something should be changed but rather pushed Jony and his designers to come up with a better solution.

As their relationship grew, Jony was also known to manage up. "A lot of times, it was Jony who would drive Steve," said Satzger. "He might say to Steve, 'I think we should change this,' if he felt that it was important to do something different." Nor was Jony afraid to go around his executive colleagues to Jobs directly if someone battled or challenged him or his team.

Jony is very protective of his design team, especially in context of the other departments at Apple. "He'll take the blame personally for screw ups," said Gautam Baksi. "He'd fall on the sword for the weakest part of the design. If it's something not up to snuff, he'd personally say it was his fault. I never felt like he threw any of the other ID members under the bus."

Although the ID team was close-knit, this collegiality did not necessarily extend to other people at Apple. A former engineer in the product design group, who worked with IDg for almost a decade, said interactions with Jony and his team were formal and strained, and he was constantly reminded of the primacy of the design group at Apple.

"I only spoke when I was spoken to," said the engineer Baksi. "These guys get paid a lot more than me and could make my life hell. I went in [the ID studio] with a purpose and left immediately afterwards. I never spent any extra time in there. I didn't socialize with the ID guys."

Jony's designers, however, often socialize together after work, especially those living in San Francisco. For many of the designers, social life and work are one. At Macworld, they often got a limo—sometimes full of Bollinger champagne—and headed out to dinner and for drinks afterward. A former product design engineer remembers being at a black-tie event at the Clift San Francisco Hotel. "Around midnight in rolls the ID crew for the after-party that is going on in the hotel lobby," the engineer said. "Stringer, Ive, Whang and a bunch more were there. . . . They're always very trendy, into trendy music."

Many of Jony's team have kids. Even though the design studio is off-limits to outsiders, the designers frequently bring their youngsters in. Satzger, for one, brought his children into the studio all the time. His daughter, who was interested in pursuing a design career, wrote a college essay about growing up in the Apple ID studio, about process and how things were built and why. "She was part of the group; she could be found there for eight to ten hours," said Satzger.

Jony was the exception as he kept his wife and twin sons, Charlie and Harry, removed from the bustle of the Apple campus. Some of the designers who live in San Francisco know his family, but to the others they are a mystery. It's an odd paradox for a man whose father had such a strong influence on his son's interest in design.

CHAPTER 8

Design of the iPod

Apple has created an art object for hardware
and software to live in.   —BONO

In the early 2000s, Apple had stabilized. The Mac was a hit, and Apple
had just introduced Mac OS X, a new operating system. OS X applications were in the works for editing video, storing photos and burning
DVDs (iMovie, iPhoto and iDVD). What was missing was an application
for digital music.

Thanks to Napster, music was rapidly turning digital and CD burners
were taking off. Apple would be almost the last computer maker to add
CD burners to its computers. In an attempt to catch up, the company
bought a third-party MP3 jukebox program for the Mac, SoundJam MP,
from a small company, Casady & Greene. Apple also hired Casady &
Greene's hotshot programmer, Jeff Robbin.

Robbin's team relocated to Apple's HQ and set about retooling
SoundJam, stripping out a lot of features to make it accessible to first-
time users. Under the direction of Jobs, Robbin spent several months
simplifying the program, which Jobs introduced as iTunes at the
Macworld Expo in January 2001.

While Robbin was working on iTunes, Jobs and the executives played
with ideas for gadgets that could be used in conjunction with the
software Apple was developing, including digital cameras and
camcorders. An MP3 player seemed a particularly obvious target, partly
because the early devices they saw on the market functioned poorly. In

the words of Greg Joswiak, Apple's vice president of hardware marketing: "The products stank."

They were two kinds of MP3 players in the market at the time. One variety was big, ugly and bricklike, based on a traditional three-inch desktop hard drive; the other, which used expensive flash memory, stored only a few songs. Neither of them worked well with iTunes, but Jobs wondered if there was an opportunity to take advantage of the intelligence built into iTunes. He asked Rubinstein to look into it.

Meanwhile, Jony's group was already making prototypes of MP3 players. The prototypes were purely experimental, nothing more than concepts for potential products, like Brunner's Juggernaut project had been. Inspired by the iMac's see-through plastic design language, the players were based on small flash memory chips, capable of storing about an album's worth of songs. "These smaller peripheral devices were considered more like extensions of a core ecosystem," said one former designer. "Engineering was not involved, at least not in these early stages of concept presentations."

Jony particularly liked an MP3 player that resembled the iMac's hockey-puck mouse, dressed out in transparent red plastic. Inspired by a yo-yo, the device had a groove around its perimeter for holding the earbud's wires, which slotted into cutouts on the back. (It looked like a round version of the earbud packaging used with the iPhone 5.) The player was controlled by a series of buttons arranged in a circle, with a small black-and-white screen in the middle. It resembled what, eventually, would be the familiar iPod scroll wheel, but at that time was purely button based with no wheel to turn. Jony's team made other versions, including some for watching video, but none of the prototypes were very compelling.

At the end of February 2001, Jobs and Rubinstein were in Japan for Macworld Tokyo. Rubinstein took a routine meeting with Toshiba Corporation, one of Apple's major component suppliers. At the end of their discussion, his hosts showed him a new hard drive, only 1.8 inches in diameter. Though tiny, it had five gigabytes of data storage—enough to hold an astounding one thousand CDs.

The Toshiba engineers didn't know what to do with the hard drive and asked Rubinstein if they should put it in a camera. Rubinstein smiled but kept his thoughts to himself. He went straight back to the hotel and told Jobs he knew how to build Apple's MP3 player. All he needed was a ten-million-dollar check.

Jobs told him to go for it, but with a catch: He wanted the new device delivered by Christmas that year, meaning Rubinstein had to have the product ready by August in order to make the marketing cycle for the crucial holiday shopping season. Ruby had six months to come up with Apple's first MP3 player.

"In Your Pocket"

As Rubinstein remembers, his biggest initial problem was that everybody at Apple, including Jony's ID group, was already busy with other products. As was usual with such exploratory, blue-sky projects, Apple went looking for an outside consultant.

Someone recommended Tony Fadell, a designer/engineer who specialized in handheld hardware and digital audio. Fadell had worked for General Magic, an Apple spinoff, and developed PDAs for Philips before launching his own start-up, Fuse Networks, in the late 1990s.

Fadell's twelve-person firm was busy trying to build an MP3 stereo player, a conventional rack-mounted component with a hard drive and CD reader instead of a tape deck or FM radio. Fadell had shopped his idea around without much success, approaching the Swiss watch giant Swatch and Palm Inc. His conversations with Real Networks brought him to Rubinstein's attention.

Rubinstein phoned Fadell on a ski slope in Aspen, Colorado, and asked him to come in and talk about a project. It was so secret, Rubinstein told him, I can't tell you what it is. Because it was Apple calling, Fadell agreed.

Only after Fadell signed a confidentiality contract did Rubinstein tell him about iTunes and their desire to build an MP3 player to hook into it. Fadell wasn't particularly enthusiastic—Apple hadn't been exactly hot in 2000—but he was flat broke. He'd burned through his start-up's cash and, because the dot-com bubble had burst, he couldn't raise any more money. Fadell took the job just so he could pay his team at Fuse.

Rubinstein offered Fadell an eight-week contract to analyze what it would take to build an MP3 player. He had to figure out the battery, screen, chips and other components, plus what kind of team would be required to get it made. When he was done, a feasibility study would be reported to Jobs.

Fadell was assigned an internal contact, hardware marketing manager Stan Ng. The two quickly formulated the design story for the new product. "'In your pocket' became the mantra for the product, because that was definitely the size and form factor that hit the sweet spot," Ng said.[1]

Fadell identified potential components, mostly from the cell phone industry, which was then rapidly taking off. Based on the size of the components he was looking at, he made mock-ups from glued-together Foam Core. A design emerged: a simple rectangle, about the size of a cigarette pack. Since it felt too light in his hand, he gave it heft by adding some old fishing weights he found in his garage. He flattened them with a hammer, then slid them between the model's foam boards.

Rubinstein loved the model and, in early April, Fadell and Ng presented it to Steve Jobs and other members of the executive team, including Rubinstein, Robbin and Schiller. Fadell hadn't met Jobs before, but he'd been coached on how to make sure Jobs picked the right design: Present three options and save the best for last. Fadell had made drawings of the first two mock-ups and brought along his Foam Core model, which he hid under a large wooden bowl that was kept on the long table in the fourth-floor executive conference room.

Ng opened with some slides about the music market and current MP3 players, but a bored Jobs kept interrupting. Fadell took over. He laid all the potential parts on the table, including Toshiba's 1.8-inch hard drive, a small piece of glass for the screen, various battery alternatives, a sample motherboard. He began talking about pricing curves of memory and hard drive storage, battery technologies and the different kinds of displays.

Then he stood up and showed the meeting a picture of his first concept, which was a big brick of a device with a slot for removable storage. Jobs said it was too complicated.

The next drawing was smaller and held thousands of songs, but it was based on volatile flash memory that would be wiped clean if the battery died. Jobs didn't like that either, so Fadell went to the table, grabbed all the parts he'd shown earlier, and began snapping them together, like a LEGO model. He handed the electronic sandwich to Jobs. As Jobs turned it over in his hands, Fadell lifted the bowl and gave him the more finished prototype.

Jobs was enthusiastic. Then Phil Schiller surprised everyone when he left the room and returned with several models for an MP3 player that featured a scroll wheel. Schiller explained that a wheel was the best way to navigate quickly through any list, whether of names and addresses or songs. The more you turned the wheel, he explained, the faster the list would scroll, making it quick to get to the bottom of a very long list. He pointed out that to select something, you hit the bull's-eye in the middle.

Schiller had gotten the idea in a meeting where he'd been examining competing MP3 players. He'd been irritated by having to hit a tiny button hundreds of times to go up and down a menu one song at a time. "You can't hit the Plus button a thousand times!" he said. "So I figured, if you can't go up, why not go around?"[2] He found that scroll wheels were actually fairly common in electronics, from scrolling mice to Palm thumb wheels. Bang & Olufsen BeoCom phones had a dial for navigating lists of phone contacts and calls that resembled the one that eventually became a signature element in the iPod's design.

Jobs asked Fadell if he could build Schiller's scroll wheel. Fadell said yes, of course. The project was code-named P-68.

Project Dulcimer

For reasons that no one seems to remember, P-68 came to be known among insiders as "Project Dulcimer." Jobs had green-lighted it, but one main player on the project, Tony Fadell, not only didn't work at Apple, he didn't particularly want to. Fadell pitched Rubinstein on awarding the job to his start-up on a contract basis, but Rubinstein refused. Instead, he extended the reluctant Fadell's contract.

As the project moved forward, however, Rubinstein got more and more uncomfortable with the arrangement: He wanted Fadell on board full time at Apple. Four weeks after the initial meeting with Jobs, Fadell was slated to present the iPod to a team of twenty-five Apple bigwigs, including Jony (who, at that stage, knew nothing of the new player's progress). With Fadell busy streamlining his ideas, Rubinstein devised an ultimatum to get him to sign on with Apple.

Before the meeting—with Jony and the others waiting—Rubinstein gave Fadell a choice. If Fadell wasn't going to take the job, Rubinstein threatened to call off the meeting. If Fadell didn't sign on to Apple on their terms, it would be the end of the project and the end of the iPod.

Fadell took the meeting and the job.

Once on staff, Fadell assumed charge of the project's engineering. Robbin headed up the software and interface team, while Rubinstein oversaw everything. The collective responsibility was obvious enough: create a product to satisfy Jobs's desire for an MP3 player worthy of the Apple marque and to do it by the looming deadline. The team's newest member, Jony Ive, would be responsible for the look, workmanship and usability of the finished product.

To meet the tight production deadline, Fadell matched up the drive from Toshiba with a cell phone battery and screen from Sony; a stereo digital-to-analog converter from a small Scottish company, Wolfson Microelectronics; a FireWire interface controller from Texas Instruments; a flash memory chip from Sharp Electronics; a power management/ battery charging chip from Linear Technologies Inc.; and an MP3 decoder and controller chip from PortalPlayer.

Fadell took a trip to Asia to meet with suppliers. He didn't tell them exactly what they might be building, but presented some vague specs for the work Apple wanted.

The earliest prototypes were built in reinforced Perspex boxes about the size of shoe boxes, which made them easy to debug. The big shoe boxes also helped disguise the fact they were working on a music player, because the team couldn't tell anyone what they were working on, even inside the company. To further obscure what they were doing, the teams put the buttons and screen in different places each time a new prototype was made. One engineer noted that it was all a silly subterfuge; a single look inside revealed immediately it was a small pocket device.

Jony's role, as he recalled later, was to help fulfill the design brief they'd been handed for Project Dulcimer. It was, in short, to create something "very, very new."[3]

Changing the Game

"From early on we wanted something that would seem so natural and so inevitable and so simple you almost wouldn't think of it as having been designed," Jony explained. The shape wasn't the issue—"it could have been shaped like a banana if we'd wanted."[4]

Given the parts of the device (the screen, the chip, the battery), the elements sandwiched naturally together into a box. "Sometimes things are really clear from the materials they are made from, and this was one of those times," said Rubinstein. "It was obvious how it was going to look when it was put together."[5]

Jony named Richard Howarth the lead designer, and they used Fadell's chunk of Foam Core as a reference. The big challenge was to design the user interface. Locating the screen was an issue, and so was whether or not to use buttons. The method of selecting songs was critical. The process inevitably reduced and reduced, resulting in a device with four buttons mounted on a dial.

Jobs worked on the interface with Tim Wasko, a veteran user interface (UI) designer who'd been at NeXT. Wasko was also working with Robbin on the UI for iTunes. He'd previously impressed Jobs with the metallic interface he created for QuickTime 4, which Jobs eventually adopted in most of Apple's software, so he was given the job of figuring out the UI for the iPod.

He started by mapping out all the options a user would face when selecting a song: the artists, their albums and finally all the songs on a particular album. "When I diagrammed it out it was a series of lists connected to each other," he said. "It was a question of pressing a button to go down to the next list, and pressing another button to come back up."[6]

Wasko created a demo in Adobe Director, a multimedia authoring program, that was pretty simple and straightforward. Before he showed it to Jobs, he replaced the original cursor keys from a keyboard with a USB jog wheel for editing video. The jog wheel had a central dial for scrubbing through video, and several buttons above and below it. Wasko drew paper labels for the four buttons on the bottom (play/pause, backward, forward and menu) and ignored the buttons on top. It worked great. Jobs was delighted with the system but pushed Wasko to get rid of the fourth button. Wasko should have known better. "If you give Steve one thing, he's going to hate it, even if it's great," remembered Wasko. "So you have to make some other crap to put on the table."

Wasko had brought nothing to sacrifice, and so he tried to find a way to get rid of one of the buttons. He labored for weeks but he just couldn't find a way to navigate the hierarchy with just three buttons. "We worked our butts off on that thing," he said.

Jobs finally acquiesced to the extra button, and Wasko took his Mac and jog wheel over to show Jony at the ID studio. "It was a quick meeting," Wasko said. "They already knew it was going to be a wheel. I just showed Jony how the interface worked."

Jony started experimenting with different places to put the screen and scroll wheel, but the options were limited. His team initially wanted to put four buttons above the wheel, just below the screen, but then decided to put the buttons around the scroll wheel instead. This made them easy to press with a thumb while turning the wheel.

"Steve Jobs made some very interesting observations very early on about how this was about navigating content," Jony would tell the *New York Times*. "It was about being very focused and not trying to do too much with the device—which would have been its complication and, therefore, its demise. The enabling features aren't obvious and evident, because the key was getting rid of stuff."[7]

To the consternation of a lot of users and reviewers, at least at first, an on-off button was omitted. The idea of pressing any button to turn the device on—and then to have it turn itself off after a period of inactivity—was a stroke of minimalist genius.

"As such a radically new product, the iPod was inherently so compelling that it seemed appropriate for the design effort to be to simplify, remove and reduce," Jony said.[8]

Other standard features of portable consumer electronics disappeared too, among them the battery compartment. Most gadgets had removable batteries, meaning they need a battery door, plus an internal wall to seal the device's guts from the user when the battery door is opened. Jony dispensed with both. A tighter, smaller product resulted, and Apple's research had already shown that no one changed their batteries anyway, even if they said they did. The sealed battery would cause an outcry, of course, because users (and reviewers especially) had come to expect a replaceable battery as a standard feature. But dispensing with it allowed the iPod's case to be just two pieces, comprising a stainless steel back, called the "canoe," which snapped into an acrylic face via an internal latching mechanism. Fewer parts also meant fewer "tolerances" (gaps) in manufacturing the product (when adjacent components are supposed to be flush, the design must allow for a tolerance; with fewer parts, alignment issues diminish).

Jony would use the same basic schema for subsequent sealed products, including several generations of iPods, the iPhone, iPad and MacBooks. "They are basically a screen and a back cover—just two parts," said Satzger. "It's a better product. A much better product."

The stainless steel back turned out to be a contentious choice: It looked great right out of the box, but was easily scratched and dented. Even if it wasn't the most obvious choice of material, the stainless steel worked, according to design consultant Chris Lefteri. "It is actually a completely irrational use of that material in that context," he said, noting that most other companies would have picked more durable plastic. "To put stainless steel on the back of a portable music player makes no logical sense, because it scratches easily, it dents, it is very heavy—but it absolutely worked."[9] One Apple executive said Jony's group chose steel simply because it was the thinnest, strongest material they could quickly work with.

Apart from the back, the iPod was "beautifully made," Lefteri said, demonstrating the design team's mastery of plastic production techniques at that point. Each iPod case was hand polished, as were the weld lines inside. "Apple did not so much use new materials," Lefteri explained, "but rather pushed the possibilities of existing ones. . . . That is Apple being very demanding of the materials, very obsessive about the levels of finish."

The white color of the iPod was Jony's idea. Jony regarded Apple's Kubrickian white stage as a reaction to the crazy color stage, which itself was a reaction to beige. "Right from the very first time, we were thinking about the product, we'd seen [the iPod] as stainless steel and white," he said. "It's just so . . . so brutally simple. It's not just a color. Supposedly neutral—but just an unmistakable, shocking neutral."[10] White also sent a message that the machine wouldn't dominate the user, unlike black tech products that tended to come off as "technical" or "nerdy."

"Shockingly neutral white" became the new normal for all of Apple's consumer products at the time. The new iMac and iBooks, as yet unreleased, were also fashioned in white plastic. "There was a whole new design language going through the shop," said the former executive. Product designer Satzger's recollections agree: "The iPod was white because the second-generation iBook was white. Most of the things Jony Ive did historically at design school back in England were white, and he started pushing white at Apple."

Initially, Jobs's instincts were against white products. Satzger at one point developed a keyboard in arctic white; when Jobs hated it, he'd presented different shades, none of them strictly white. His range of whites for plastic materials included shades he called cloud white, snow white, and glacial white. Another was moon gray, which appeared white but was actually gray. When Satzger showed Jobs a moon gray chip, he could offer reassuringly, "It's not white." A sly move, as Jobs approved the moon gray keyboard. Likewise, the iconic iPod headphone cables weren't white, but moon gray. "Moon gray and seashell gray were shades developed by us at Apple that were so close to white as to appear almost white, but were in fact gray," explained Satzger.

The white plastic face of the iPod was coated with a very thin, transparent Perspex layer that gave it its sheen. The transparent layer was raised so little above the iPod's front that it could only be seen when held sideways on. This was the iPod's clear, sealed lid. The clear coating also put "quite a strong, almost a halo around the product," said Jony.[11] It dazzled.

With the iPod coming together, everyone got more and more excited. Steve Jobs was working almost every day with Robbin and Wasko on iTunes and the UI, while Jony and his team were occupied with perfecting the ID. The pressure to deliver was high but, even at the time, the team sensed they were making something remarkable. "The design of the iPod turned into a really personal project," said Jony. "I mean, I love music. The team loves music. And I think collectively, it's a product we're looking forward to getting our hands on."[12]

New prototypes had to be delivered on Fridays, which was unusual, because prototypes were usually on hand in the middle of the week so that they could be worked on during rest of the week. Though Jobs's daily involvement wasn't widely known because the project was siloed, some members of the Dulcimer team suspected that Jobs was taking the prototypes home on Fridays, and playing with them over the weekend.

One reason for such suspicions was the rash of new demands that arrived on many a Monday.

Prior to launch of the new player, nothing was left out of consideration, including the new product's packaging. The packaging became almost as important as the overall product design. Previously, boxes were designed primarily for shipping, but with the iPod, the design team focused on the customer, rather than the transport company. The decision was made to design separate shipping containers and retail boxes so the customer wouldn't be taking home his or her iPod in a plain-Jane shipping box. The result was an elaborate box that cradled the iPod like a piece of jewelry. "The iPod was the first product where we thought about the packaging as almost as important as the overall product design," Satzger explained. "Packaging, it's just as important as everything else."

In August, one of the physical iPod prototypes finally played a song. A group of people working late that night took turns listening to music on the new gadget, hooked up to headphones from someone's old Sony Walkman. That first song was Spiller's "Groovejet (If This Ain't Love)," a house-music dance tune with vocals by the British diva Sophie Ellis-Bextor.

"Oh, my God," Jobs said. "This is gonna be so cool."[13]

The Unveiling

"We have something really exciting for you today," said Steve Jobs on October, 23, 2001, at a special press event on Apple's campus. Jobs had asked only a few dozen journalists to a product unveiling. The invite said simply, "Hint: It's not a Mac." Just a month after the September 11 attacks, and the world was still shell-shocked and, by Jobs's showman standards, the event was low-key.

When Jobs first pulled the iPod from his jeans pocket, the reaction from the audience was muted. It didn't seem that exciting, especially when the audience learned of its price: $499. Nearly $500 for an MP3 player—and one that worked only on the Mac, not Windows—seemed unrealistically high. Early reviewers were just as skeptical, with one saying that iPod stood for "Idiots Price Our Devices."[14] The iPod sold only modestly at first and didn't take off until two years later, when it was made fully compatible with Windows. Still, the seeds of the iPod's success were sown with the first device and Jony was confident in the new product.

"Our goal was to design the very, very best MP3 player we could; to design something that could become an icon," Jony said in the iPod's first promotional video.[15]

Looking back on the process, Jobs believed the creation of the iPod was quintessential Apple. "If there was ever a product that catalyzed Apple's reason for being, it's this," he said, "because it combines Apple's incredible technology base with Apple's legendary ease of use with Apple's awesome design. Those three things come together in this, and it's like, that's what we do. So if anybody was ever wondering why Apple is on the earth, I would hold up this as a good example."[16]

Yet it was also an odd-duck project, led by engineering, not Jony's design group as most of the products up to that time had been. Because of the rush to market, it was assembled from off-the-shelf parts and Jony was brought in to do a hated "skin job." Yet he managed to put his mark on it by making it white, the color he'd championed for high-tech products since he was in college.

The project also earned him the nickname "Jony iPod," and launched an armada of white tech products. The iPod would do for white what the iMac had done for translucent plastic. And Jony accomplished this tidal shift against the wishes of Steve Jobs, who had initially resisted white products.

The iPod introduced numerous design features that would be used to dramatic effect in subsequent products, including Apple's first touch interface (albeit a simple one). The iPod set the standard for many later products with its sealed case, compact design and radical ease of use— all Jony's team's work. It was also Apple's first mobile product in the Jony/Jobs era; its development allowed Jony's team to perfect the design and manufacture of portable products, thereby setting the standard for seamless cases and sealed batteries that eventually the whole industry adopted.

The iPod stands as a remarkable accomplishment. U2's Bono put its charm rather neatly when he said of the iPod, "It's sexy." Another adjective applies too: ubiquitous. The iPod's appeal soon made the device a phenomenon.

"[The iPod] was the first cultural icon of the 21st century," said Dr. Michael Bull, a lecturer at the University of Sussex whose studies have earned him the nickname "Professor iPod." "Roland Barthes argued that, in medieval society, cathedrals were the iconic form. Then, by the 1950s, it had become the car. . . . I argue that 50 years later, it was the iPod, this technology that let you fit your whole world into your pocket. It was representative of a key moment in the social world of the 21st century."[17]

CHAPTER 9

Manufacturing, Materials and Other Matters

One of the great things about our team, about working so closely together is the feeling that we're really only at the beginning of something, that we have only just started. We still have lots more to do.
—JONY IVE

Over the years, the genius of Jony Ive's IDg studio has been most apparent when the team confronted special challenges. A blending of epiphany thinking and practical implementation became a signal characteristic of the Jobs-and-Jony collaboration. More often than not, the original solutions the team came up with pushed the boundaries of traditional manufacturing; the refining of the design of the original iMac is a case in point.

About eighteen months after the first iMac hit the market, Jony's team starting to think about replacing its bulky CRT with a light, thin LCD screen. The team began work in 2000, and the project proved challenging, requiring scores of prototypes. But the finished product would be one of Apple's most distinctive computers.

At first, Jony's team came up with a conventional concept for a flat-screen computer, with the guts of the computer attached to the back, much like the earlier Twentieth Anniversary Mac Jony had designed. But Steve Jobs disliked what he saw as ugly and inelegant.

"Why have this flat display if you're going to glom all this stuff on its back?" he asked Jony. "We should let each element be true to itself."[1]

According to Walter Isaacson's biography, Jobs left Apple HQ early to think it over at his home in Palo Alto. Jony dropped by and they took a walk in Jobs's garden, which Jobs's wife, Laurene, had planted with sunflowers. Jony and Steve were riffing on their design problem, when Jony wondered what the iMac might look like if the screen was separated from the other components like a sunflower on its stalk.

Jony became excited and started sketching ideas. "Ive liked his designs to suggest a narrative," wrote Isaacson, "and he realized that a sunflower shape would convey that the flat screen was so fluid and responsive that it could reach for the sun."[2]

A former executive tells the story differently. Jony made two prototypes. One was the ugly, inelegant flat-screen unit, the other a "goose neck" design with a separate screen and base. At the presentation, Jobs chose the gooseneck design because it was "anthropomorphic." Like the original Mac, Jobs wanted a "friendly" computer.

Jony's team next faced the problem of attaching the screen to the base.

First they tried a series of ball-and-socket pieces that resembled the vertebrae of a spine. The vertebrae were held together by a system of spring-loaded cables with a clamp attached to the back of the screen. When the clamp was tight, it tensioned the cables and held the vertebrae in place. When the clamp was loosened, it put slack on the cables and allowed the spine to be moved. "The display floated so that you had to grab it with two hands to release the lock, and then when you positioned it, it locked in place. It had a series of these beautiful balls and sockets, and the power and signal cables went through the neck," said Satzger. "When you released everything it would relax, and when you tightened it, it would lock with a big cam mechanism."

The team made scores of prototypes, which might have looked beautiful, but they turned out to be impractical. Locking and unlocking the spine clamp took two hands, making it difficult for some users, especially children, to adjust the monitor.

Temporarily stumped, Jony asked the design consultancy IDEO to come in and take a look. IDEO was supposed to evaluate the usefulness of the design, but instead their designers suggested replacing the spine with two rigid arms that resembled an anglepoise lamp. That seemed like a great idea, and much more practical.

Jony's group made several more prototypes and found that IDEO's two-segment arm worked well. But Satzger wondered aloud in one of the brainstorming sessions, "Why do we need that much flexibility? Why don't we adopt a one-arm mechanism?" Satzger's suggestion led nowhere—until Jony and Steve returned to the IDg studio after a meeting. Steve also suggested dropping the second arm.

Again, Jony's team set to work. After a lot of engineering, the arm they came up with was of stainless steel that, thanks to an internal spring under very high pressure, balanced the weight of the screen perfectly. The screen could be moved easily with just one finger; its cables ran internally.

"We were all really excited about it," Satzger said. "We loved it. We learned a ton of things too doing it." Jony summed up their accomplishment, calling it "an engineering tour de force. That was an extremely difficult problem to solve. . . . [The arm] seems simple, but that simplicity belies something very complex."[3]

Jony's team also agonized over the bezel, the plastic frame around the display. The earliest prototypes had a very narrow bezel but the designers found that, when adjusting the display, it was almost impossible not to poke the screen, producing that rippling effect that reminds you what the "L" in LCD stands for (liquid crystal display). When they tried a thicker bezel, Jony thought it "took away from the story of this amazingly bright, light display."[4]

That led to the idea of a "halo," a wide rim of transparent plastic that would give users something to grab without ruining the aesthetics of the screen. The halo was used to great effect in the iPod, and became one of the most recognizable design motifs of Jony's clear-on-white era, one that would carry through to the bezel of the present iPad.

The dome base of the iMac was another feat of engineering. The iMac G4 crammed a computer, drives, and a power supply into its hemispherical base. It had a cooling system borrowed from the Cube, which sucked air in through the bottom and expelled it out of the top, but, unlike the Cube, the chips ran hot, requiring a fan. Even so, Jony remembered, "There's not a screw or detail that's not there for a very good reason."[5]

According to Jony, the design of the iMac G4 was ingenious not because of its shape but its unexpected unobtrusiveness. Although it looked like a freaky lamp on approach, everything but the screen disappeared when the user sat in front of it. "With the new iMac, if you just sit there for ten minutes and move the display around, you quickly forget about its design. The design gets out of the way," Jony concluded. "We are not interested in design statements. We do everything we can to simplify design."[6]

As with the marketing of the iPod, Jony's team also designed the iMac's packaging. Boxes may seem trivial, but Jony's team felt that unpacking a product greatly influenced the all-important first impressions. "Steve and I spend a lot of time on the packaging," Jony said then. "I love the process of unpacking something. You design a ritual of unpacking to make the product feel special. Packaging can be theater, it can create a story."[7]

Though they took the process seriously, that didn't mean they lacked a sense of humor. As a joke, the design team designed the inside of the iMac G4's box to look like male genitals. "You had the neck laying there and the two ball speakers next to it," said Satzger. "People would open the box and say 'What?'"

When Steve Jobs unveiled the iMac G4 at Macworld in January 2002, *Time* magazine featured it on the cover that week, just the second time a product launch had made the magazine's cover.

"This is the best thing I think we've ever done," Jobs said on stage as he introduced the machine. "It has a rare beauty and grace that is going to last the next decade."

When he showed a picture of the base, he said, "Isn't that the most beautiful bottom of a computer you've ever seen?"

Jony appeared in a promotional video that Jobs showed to the crowd. "The easy part was knowing we were going to use a flat panel display," Jony said to the camera.[8] "The hard part was how. Our solution appears to defy gravity. It's just this very simple, pure frame that appears to float in space. When you look at it now, it seems so simple, it seems so obvious, and yet again, the simplest, most efficient solution has been the most elusive."

Following Jobs's keynote, Jony quietly walked the MacWorld show floor, trying to gauge people's reactions. He'd worked on the various pieces of the iMac in secret for two years, and, with no public feedback, Jony worried that the marketplace might not like the imaginative new machine. "I think they like it," he concluded later. "Yeah, yeah, people are . . . pretty enthusiastic."[9]

He needn't have worried. The iMac G4 helped reset Apple's public image, which had suffered after the failure of the Cube. With Jony at the design helm Apple was at the top of the game again, with Jony helping set its course.

Jony at Work and Play

His role at Apple assured, Jony indulged his old passion for cars. He treated himself to an Aston Martin DB9, a supercar known for its association with James Bond. Jony had the car delivered to New York and drove it cross-country with his dad, Mike. It cost about $250,000, but just a month after he got it, Jony wrecked the car on Interstate 280 near San Bruno. The accident nearly killed him and his commuting partner, Daniele De Iuliis, who was riding in the passenger seat.

"Jony was going pretty fast, although he said he was not going over eighty miles per hour," said a colleague. "Something happened in the traffic. Jony lost control of the car, which went into a spin. It sling-shotted the back end, whacked into a panel truck and knocked that over, and went straight into the median. The whole car was smashed. They were lucky to get out alive. The car was a mess; totally fucked up on all sides."

The car's airbags went off, filling the car with the smell of the explosive that set off the airbags. Jony found the smell unsettling as he came to. "He woke up with the smell of gunpowder in the car and that was weird. He was distressed by that," said another source. "Ironically, the car crash alerted Apple to how important Jony is to the company, and they gave him a big pay rise."

Jony was undeterred in his quest for speed and cool cars: He bought a second DB9. When it burst into flames parked outside his garage, he complained to Aston Martin. "Him being English and his relationship with Steve and Apple, he went to Aston Martin and they told him they'd give him a great deal," said a source.

The company offered him a discount to move up to the Vanquish (2004–2005 model), a $300,000 grand touring car with a monstrous V12 engine. Soon after, Jony bought a white Bentley, another powerful British luxury car. He also purchased a Land Rover LR3 after one of his colleagues in the design studio bought one. "Jony wanted one as well and got one within days," said a source. Later, Jony added a black Bentley Brooklands to his stable. Costing about $160,000, the Brooklands was hand assembled with lots of interior wood and leather. It's another powerful machine, capable of reaching sixty miles per hour from a standing stop in five seconds.

As well as being fast and powerful, Aston Martins are known for their innovative production methods. Their cars are built from unusual, lightweight materials like aluminum, magnesium and carbon fiber. The all-aluminum chassis is glued together rather than welded, which makes it incredibly strong and resistant to cracking. Jony would soon introduce similar production methods to Apple's manufacturing arsenal.

Starting with the iMac G3, Jobs and Jony worked together more closely than ever. Ken Segall, the TBWA/Chiat/Day ad man, who continued to consult with Apple, reported that "[Jony] was in most of our biweekly meetings with Steve." These were marketing meetings—not Jony's thing—yet Jobs liked him there to bounce ideas off. "Steve obviously valued Jony's thinking for more than just product design."

His role as consigliere was cemented by their relationship outside the conference room as well. Segall also remembered: "Steve and Jony having lunch in the cafeteria—and not once in all that time did I ever see Steve there without Jony. They seemed truly inseparable."[10] Meanwhile, Jony's relationship with Rubinstein was worsening. They fought constantly over everything.

In the wake of the first iPod release, it became evident that Jony's role in shaping Apple's philosophy had grown. His belief that computers and music players should be simple to use and beautiful to look at drove many decisions not only in the evolution of the iPod but in the release of new models of the iMac and iBook.

"Apple puts out something really elegant like the [iPod]. Then they relentlessly improve it," said Dennis Boyle, one of the cofounders of IDEO. "So not only are they outstanding at putting true innovations onto the market, but also at making those products better and better. . . . They leave their competitors in the dust."[11] Within two years of launch, the iPod was made Windows compatible (it would have been quicker, but acquiescing to Windows was a big psychological barrier for Jobs), and gained the iTunes Music Store, which made it an easy matter to load the player with new content.

In continuing the process of miniaturization, which had shrunk all of the iPod's components, the iPod mini hit the market in January 2004 with a smaller, solid-state click wheel that was touch-sensitive. The buttons at the four compass points were incorporated into the wheel itself. "The click wheel was designed out of necessity for the mini because there wasn't enough room for [the buttons on] the full size iPod," said Jobs. "But the minute we experienced it, we just thought, 'Oh my God! Why didn't we think of this sooner?'"[12]

Jony offered a more detailed version of the mini's development. Originally conceived as a small iPod, the first versions, which used the same materials and design language, weren't working. "It was just completely wrong," Jony said. "Then we started to explore very different materials and approaches. We realized we could make this in aluminum. Unlike with stainless steel, you could blast it and then anodize it— which is a form of dyeing—and then you could do color in an unusual way."[13]

That first foray into aluminum would influence a whole generation of products. Like the iMac before it, the iPod mini would come in a range of colors. It was a big hit; the fastest-selling iPod up to that time, especially with women. It was the first iPod that people started wearing on their bodies, outside their pockets, with a strap or a clip. Some treated it like an accessory, a piece of fashion jewelry. The mini also kick-started the trend of having a small, dedicated iPod just for the gym or running.

In just four years, Apple took the iPod from the 6.4 oz. original to the 4.8 oz. nano. In the process, storage was increased sixfold, a color screen and video playback were added and battery life was extended to four hours. And the price was reduced by $100. Eventually, Apple was selling a player at every $50 price point between $50 and $550, including the shuffle, which dispensed with the screen, an exercise in ballsy minimalism that perhaps only Jony and Jobs could have pulled off.

Some of the gains were a function of manufacturing improvements. In an interview with the British edition of GQ, Jony expounded upon the advances he'd achieved with the first aluminum shuffle. Machined from extruded aluminum, the shuffle clipped together with barely a gap between the parts. "The way the parts fit together is extraordinarily tight," Jony said. "I don't think there's ever been a product produced in such volume at that price, which has been given so much time and care."[14]

The world was taking notice not only of Apple's products but of the company's talented head designer. Jony had been winning prizes and awards since his teens, but in the early 2000s, the awards really poured in. In July 2002, the Industrial Designers Society of America honored Jony and Apple with the design world's highest honors, the gold Industrial Design Excellence Award (IDEA) for the original iPod, declaring the music player "the most memorable design solution" of the year.[15]

In June 2003, the London Design Museum announced that Jony was the winner of its inaugural Designer of the Year Award. His prize was £25,000, plus a golden gong. The distinction seemed a foregone conclusion. "Designer of the Year doesn't really come close to describing what Ive has achieved in the decade since he joined Apple," wrote Marcus

Fairs in *ICON* magazine. "There are few designers who have had the commercial, critical and sociological impact of Ive and his small team at Apple."[16]

Jony took every opportunity to include his colleagues at awards ceremonies, a clear acknowledgment that the work he was being celebrated for was always a group effort. At the London Design Museum he was joined at the party by his design group, who paid him the compliment of dressing "exactly like Ive, had crewcuts like Ive, and said as little as Ive," said *ICON*. Jony told reporters he found awards "nice" but "slightly awkward in that it's a hard thing to receive when there's a team of you."[17]

"One of the great things about our team, about working so closely together," Jony added, "is the feeling that we're really only at the beginning of something, that we have only just started. We still have lots more to do."[18]

Jony's and Apple's success was quite obvious. As he noted on a trip to London, everywhere he looked, he saw the white earbuds. Having labored for years when Apple had just a small share of the computer market, he found it gratifying to see one of his designs become dominant.

Jonathan Glancey, design critic for the *Guardian*, said Jony's genius had been "to make imaginative what was previously lackluster, to give a glamorous, desirable and human face to a technology that has been . . . the domain of joyless office managers and electronic professionals (transforming) a humourless technology into something desirable and sophisticated."[19]

Canada's *Saturday Post* called the iPod "the defining device of this generation's iWant-iNeed-iWish gadgetophiles." Its ubiquity and "no beige please, we're British" design philosophy granted it permanent icon status.[20]

Toward the end of the year, Jony's growing reputation was recognized by the government of his native England. On the heels of the cultural era known as "Cool Britannia," Prime Minister Gordon Brown called out Jony as a model for English design innovation. The *Guardian* reported that Brown sought to use the nation's large number of design graduates "to make [UK] products more desirable than low-cost competition, and thus fend off the challenge from China and India."[21]

In the mid-nineties, according to the report, one in sixty-four graduates were in design programs. A decade later, the number was one in sixteen. "Design is not incidental to modern economies but integral; not a part of success but the heart of success; and not a sideshow but the centrepiece," Brown said. His government commissioned official reports to find the economic potential of design in its industries and found that design-centered companies "saw a turnover rise by fourteen percent and profits by nine percent."[22] No doubt Jony's father, Mike, had made a huge contribution to the rise of design in his native land and he was honored thusly. In 1999, in recognition of his contributions to British design education, Mike Ive was awarded the Order of the British Empire (OBE).

In 2003, Jony was appointed a member of Royal Designers for Industry; in 2004, he was awarded the RSA Benjamin Franklin Medal; and in 2005, he won what was to be the first in a string of prestigious awards from the British Design & Art Direction (D&AD). In 2006, he was named Commander of the Order of the British Empire (a higher award than his father's OBE).

Jony didn't make a public comment about the award at the time, but in a statement, Apple said: "We are as proud as could be that Jony is receiving such a prestigious commendation."[23]

Although his designs were drawing much notice, bigger work (quite literally) was still ahead. The year 2003 saw the release of the seventeen-inch PowerBook. It was a monster laptop, but none of Apple's marketing materials mentioned Jony's proudest innovations, which included its internal frame and a clutch mechanism in the lid hinge.

The variable-rate clutch he devised had less resistance in the near-closed position, allowing the lid to be opened with one hand without the bottom of the laptop lifting off the desk. A consuming attention to detail contributed to the user experience, even though few users have any idea how much work went into it.

Jony was proud of the PowerBook's construction, and dismantled one for his 2003 Designer of the Year exhibition at the Design Museum. "We took [it] to pieces so you can see our preoccupation with a part of the product that you'll never see," Jony said. "I think—I hope—there's an inherent beauty in the internal architecture of the product and the way we're fabricating the product: laser-welding different gauges of aluminium together and so on. Very often people assume that it's only if it's a smaller volume production—batch production—that people will really be caring about all of the details. I think one thing that is typical about our work at Apple is caring about the smallest details. I think sometimes that's seen as more of a craft activity than a mass-production one. But I think that's very important."[24]

Richard Powell, founder and director of the famous design firm Seymourpowell, liked what he saw. "When you talk to Jony Ive, his eyes sparkle with the memory of a design challenge overcome, a problem solved, a material found. He becomes animated about a surface perfected and a process explored. For Ive, nothing is left to chance; everything must be deeply considered."[25]

Powell saw Jony's focus as defining. "Innovation," he wrote, "is rarely about a big idea; more usually it's about a series of small ideas brought together in a new and better way. Jony's fanatical drive for excellence is, I think, most evident in the stuff beyond the obvious; the stuff you perhaps don't notice that much, but which makes a difference to how you interact with the product, how you feel about it."

The iPod was becoming a monster hit, suddenly as important to Apple as the Mac line. In 2004, the iPod was spun off into a separate division and Rubinstein, formerly head of all hardware, put in charge. In executive meetings, Jobs and his team started wondering what else the company might get into. An Apple-branded car and digital cameras were some of the ideas knocked around.[26]

In 2005, Jobs promoted Jony to senior vice president of industrial design, elevating him to the same senior level as Rubinstein. Jony had reported to Rubinstein, with whom he fought constantly. Now Jony answered only to Jobs.

Jony and Ruby had been getting into regular shouting matches. Jony was always pushing the envelope, constantly challenging how things were made and designed. It was Rubinstein's job to get products out the door, and he frequently balked at Jony's demands. According to a former designer who worked with the pair, Rubinstein avoided Jony and the studio as much as possible and, if he did have to meet with Jony, he became visibly agitated. "Jon's blood would just boil whenever he had to go into the studio and deal with Jony," said one source.

The relationship was equally problematic for Jony. The confrontation that had been brewing for years finally happened. Jony reportedly went to Jobs and told him, "It's him or me."

Despite Rubinstein's essential role in the development of the iPod and scores of other products, Jobs chose Jony.[27] In October 2005, Apple issued a press release that framed Rubinstein's exit as a long-deserved retirement. He was replaced as head of the iPod division by Tony Fadell.[28] Rubinstein would spend some time building a house in Mexico before later becoming CEO of Palm and developing a rival to the iPhone.

Speaking of the affair years later, Rubinstein was diplomatic about his relationship with Jony. "Jony and I worked very closely over many years and did a lot of work together. My job was to instill balance and get products out the door. Sometimes working with Jony could be difficult."[29]

Even if they had their hard moments, during their shared time at Apple, Jony and his then-superior Rubinstein steered the company's design language through several phases, moving from multicolored plastic to monochrome plastic and on to a range of metals. Importantly, each stage marked a growing sophistication in the tandem of design and manufacturing.

That the manufacturing methods became a larger and larger part of the design process undoubtedly added to the tension between the two men. The design team was no longer concerned with merely how the products looked and worked. Their brief became how the products were made, and Jony's team, which had always spent a lot of time designing products, invested more and more time in figuring out how to manufacture them.

Former design team leader Bob Brunner gave his view on the travails: "Apple designers spend ten percent of their time doing traditional industrial design: coming up with ideas, drawing, making models, brainstorming. They spend ninety percent of their time working with manufacturing, figuring out how to implement their ideas."[30] It's little wonder, then, than Jony's star was ascendant in a company where design and materials had become the industrial equivalent of conjoined twins.

Jony's outlook on the cost of R&D and development work was simple: He didn't want to know about it. As he told a former engineer who worked in the operations group, "I do not want any of my guys thinking about cost. They should not even care about the cost because that is not their job."

To some in the company, it seemed as if Jony reported to no one, not even Jobs. Reportedly he told suppliers, "Imagine I have a bucket of money in my hand. I will let you pull out as much as you want to make this happen," said the operations engineer. Discussions between Jony's group and product development and operations tended to go in one direction, from the designers to production.

As the operations executive summed it up, "ID rules Apple."

Streamlining Manufacturing

The intimacy of design and manufacturing is what led Apple to China. Apple's shift to manufacturing its products in China has been credited to Tim Cook, the company's CEO and Jobs's chosen successor.

Jobs himself had been managing Apple's suppliers and factories when he hired Cook in 1998 as senior vice president of operations. Raised in Robertsdale, Alabama, Cook was formerly an operations executive at Compaq and had spent twelve years at IBM. With his cool demeanor, he bonded immediately with the volatile Jobs, who had already rejected a number of candidates for the operations manager role, walking out of at least one interview within five minutes. But the formidable CEO connected with Cook and offered him the job, giving him an office near his own.

Cook inherited the unenviable task of overhauling Apple's manufacturing and distribution networks, which were in a notorious state of disorganization. Apple had its own factories operating in California (Sacramento), Ireland (Cork) and Singapore. The three sites produced motherboards and assembled the same products that, in theory, were intended for sale in their respective geographies, America, Europe and Asia. In practice, however, motherboards were often shipped from Singapore to Cork for partial assembly, then shipped back to

Singapore for final assembly, and then to the United States for sale. "As you can imagine," Cook said of the arrangement, "the costs weren't so good, and the cycle times weren't that good."[31]

Jobs's streamlining of Apple's product line to just four products simplified things. Instead of four motherboards for desktops, there was now just one. The machines in his 2×2 product matrix shared as many common parts as possible, and instead of using exotic Mac-only technology, they used industry standard parts shared by other PC manufacturers.

Still, Apple's factories were costly and inefficient, so Cook started using outside manufacturers. Cook began by visiting every supplier that did business with Apple. He struck hard bargains, consolidated suppliers and encouraged suppliers to move their own factories close to assembly plants.

When Apple introduced the iMac in 1998, it was initially made at the three Apple factories, though LG made the iMac cases and monitors. In February 1999, the company shifted streams, outsourcing the iMac entirely to LG and selling off Apple's factories. In 2000, Hon Hai Precision Industry Co. Ltd. took over iMac production. The electronics manufacturer, based in Taiwan, is better known internationally as Foxconn.

Cook would do the same thing with laptops, shifting production from Apple factories to Quanta Computer in Taiwan (for the PowerBook) and Alpha Top Corporation in Taiwan (the iBook). By moving production to outside partners, Cook solved one of Apple's biggest headaches: inventory, namely, stock sitting in storage. The more parts and machines Apple had in its warehouses, the more money the inventory cost the company. Warehouses full of unsold machines had nearly sunk Apple in 1996, so the new normal had become the less inventory, the better. Cook once referred to inventory as "not only evil, but fundamentally evil."[32]

Too much inventory was a consequence of forecasting sales in advance, which had always been mostly a matter of guesswork. Traditionally a company would manufacture goods to suit expected orders over a period of months; that also meant that each machine that was built, shipped and stored cost money until it was sold.

Cook wanted a better system, looking to harness emerging information technology programs that made it possible to fulfill actual customer needs. He created a state-of-the-art IT system that allowed Apple to build in response to demand. He helped set up a complex enterprise-resource-planning system (ERP). The Intranet-based system hooked directly into the IT systems at Apple's parts suppliers, manufacturers and resellers, giving Cook a detailed view of Apple's entire supply chain, from screws to customers. With that data, he could manage daily production on the basis of weekly sales forecasts and keep precise tabs on stock at its resellers. He could tell if CompUSA had excess inventory or if it was running out of stock. Later, the ERP was extended into Apple's own retail stores and became so precise it tracked and reported sales every four minutes.

The ERP allowed Apple to build computers only if they were needed, so-called just-in-time production. And it allowed parts to be left in suppliers' warehouses until they were required.

Within seven months of Cook's arrival, Apple had reduced its on-hand inventory from thirty days to six. By 1999, inventory had been reduced to just two days, beating by far the industry's gold standard, Dell. Thanks to improved operations, Cook was credited with playing a big part in stemming Apple's losses and returning it to profitability.

Over the years, Cook fine-tuned the system until it was capable of delivering millions of products in secret just in time for massive product launches, accounting for much of Apple's massive growth. In overseeing Apple production lines, Cook successfully managed not only to keep inventory low but to keep profit margins high at Apple. The company could never have grown so rapidly and so large without such operational excellence. Just as Jony and the design team designed great products, Cook and his team figured out how to produce them in their millions—and deliver them all over the world, on time and in utmost secrecy.

Aloo-MIN-ee-um

Another reason Apple shifted manufacturing to China was that the design team started designing products in aluminum, and that's where the supply chain was located. The Titanium PowerBook G4 had been a big hit, but titanium is an expensive and difficult metal to work with. Plus, it had to be coated in a metallic paint to protect against scratches and fingerprints, but the paint had a tendency to flake off. When Jony's first prototypes for the iPod mini didn't work in acrylic and steel, he shifted to anodized aluminum.

According to his design team's research, aluminum looked like a good material for laptop cases and iPods. It is strong and light, and it can be finished in a range of colors when an anodized coating is bonded chemically to the metal. At that stage, Jony and his crew knew little about aluminum manufacturing, so they started researching camera manufacturers like Sony, which produced a lot of cameras in aluminum.

These Japanese-based manufacturers turned out elegant, long-lasting and well-made products. But the aluminum's point of origin was in China. "We got introduced into that supply chain," said Satzger. "I remember many trips to go over to understand how you do things with aluminum."

Apple has taken much flak for outsourcing to China, but when Jony and the team first started using aluminum, they tried to work with manufacturers stateside. Satzger, who was in charge of materials and finishes for Jony's group, made the initial contacts. He researched the suppliers to find companies that could make components in the quality and quantity that Apple required.

When the design team was creating the first Mac mini, Satzger started working with a U.S.-based aluminum supplier. The directive from Tim Cook's operations group was clear: They wanted the Mac mini to be manufactured in the United States. The U.S. supplier looked like a good bet because it was able to supply high-quality aluminum that was relatively free of impurities and would anodize well.

The Mac mini looked relatively simple, but its case was surprisingly complex. The mini's square case was made from sheet aluminum extruded into a square shape, which was then machined to achieve the right tolerances and finish, especially on the top. Then the case had to be anodized. Again, the design group had exacting requirements, as the anodized layer had to be just the right texture, color, gloss and thickness.

Satzger spent months working with the U.S. supplier but as the deadline approached, the company failed to produce sample cases. A stressed-out Satzger went to the operations group. "We reached a point in the program where I said to the representative of the operations team that was pushing the move toward manufacturing in the USA, 'We have a really tight schedule, and the [US manufacturer has] not yet delivered a part in spec from the extrusion machine; they haven't delivered a finished part. When are they going to do this? If they don't deliver, we don't have a product. Where is our backup plan?'"

There was none. An increasingly frustrated Satzger observed that the American supplier had no concept of the quality that was required by Apple. "At Apple, good enough is not good enough," said Satzger. "American companies could not understand the quality that is required for a customer-facing part on an Apple product—the little things that customers notice."[33]

By contrast, when Jony's team went overseas for other projects, the Asian suppliers bent over backward to get the contract. "We first went to Japan, and started using titanium, and moved from there onto aluminum for the first iPods and PowerBooks," remembered Satzger. "Then we were able to take that knowledge and experience and move, along with a couple Japanese companies, into China and we could say to people such as Foxconn, which had been molding parts for us, 'Can you make basic sheet metal for us?' and we started working with them from there. In China, their attitude was very much, 'We're going to work on a product to make sure it meets all its specs.'"[34]

The Mac mini would be produced in Asia, along with iPod minis and other products, at Foxconn.

Another product, the Power Mac G5 tower, would mark a big shift in the relationship between Apple and Foxconn. The design group wanted to make the tower in aluminum instead of plastic, as it had done with its predecessor, the Power Mac G4. Even by Apple's standards, the project was immensely challenging. "Multiple people got shifted [to different jobs] because they just couldn't do it," said Satzger.

The project took more than a year. One cause of the delays was the 2003 outbreak of SARS, which quarantined some of the design team at Foxconn, including Jony. "I stayed for three months in a dormitory to work on the process," he said. "Ruby and others said it would be impossible, but I wanted to do it because Steve and I felt that the anodized aluminum had a real integrity to it."[35]

Jobs wanted the case to be an extruded part, and he thought the big swooping handles on the previous towers were really important to the image of the product. Jony's team started looking for a huge extruded aluminum tube that could be flattened like a lozenge, and then have two holes scooped out to create the handles. But they found that the only extruded aluminum tubes at the time were eighteen-inch tubes made for plumbing, which would be too small to suit their needs. Jony's team investigated whether they could take two extrusions and join them together.

As they struggled, Satzger suggested that instead of using extrusions, the case could be done by the process of roll forming, the way steel gutters are made. A flat sheet of aluminum could be bent at various points as it is passed through a series of rollers to make the lozenge shape. When Satzger suggested this in one of the group's biweekly brainstorming meetings, one of the other designers said it wasn't a good idea because Jobs had set his mind on an extruded case.

"You don't get it," the other designer told Satzger. "Steve wants it extruded."

"No, you don't get it," Satzger replied, "it's not possible."

Finally, Satzger went to Jobs and persuaded him to go with the roll process, despite the opposition of the whole team. The sheets were roll formed into a C shape, with a large door installed on the open side. At first, Jony's team was concerned about the two small joints on the open side (they wanted it seamless), but eventually decided they could live with it because the joints couldn't be seen from the front.

Because customers would be opening the machine up, Jony decided they had to design the internal components too. "That was the first time we went internal," said Satzger. "We controlled board color [the color of the motherboard]; we controlled every connector, every cable. We designed every part inside it: the fan housing, the plenum for the air flow."

Jony's team spent months trying to find a good way to lock the door in place. Their first efforts were elaborate locking mechanisms on the door itself, but those ruined the case's austere surface. Next they decided the G4 tower would have a locking ring, like the recessed deck latch on a sailboat. In one of the brainstorms, Satzger suggested putting the latch on the back of the machine. "I finally said, 'Why does it have to be on the door? Look at the hood latch on your car. Don't interrupt a surface with other details. Leave it for the Apple logo.'"

The team designed a latch that resembles an automobile hood latch connected to a pair of slender deadbolts. The latch is indeed on the back of the machine and, when activated, it slides deadbolts on the insides of the door at the top and bottom that hold it in place. It's complex but elegant, with no sign of the mechanism from the outside.

When the first cases came off the line, Jony saw that the new machine would become a showpiece. Machines in this category were usually put under a desk, but this one looked so good, Jony figured, it would be put on top of desks instead. That meant all the surfaces had to be treated like the front.

With most products, the front is what's called an "A" surface. As the best surface, it has to be finished to the highest standards. The sides are "B" surfaces, the backs "C," the insides "D" surfaces. But, said Satzger, "This product was so beautiful, everything was an 'A' surface."

When the design team conveyed this directive to the manufacturer, Foxconn was flabbergasted. "They were like, 'What do you mean? This is crazy,'" recalled Satzger. "It got to the point where Foxconn's team said we've never done anything like this, we just can't." In the end, they would meet Apple's standards.

Offshore

For better and for worse, Apple has become the poster child for the ills of offshore manufacturing. Foxconn assembly plants in particular have attracted criticism; after a rash of worker suicides in 2009, negative international attention resulted in investigations that exposed a host of labor abuses.

Foxconn had as many as half a million workers in some of its plants, assembling iPhones and iPads by hand. The workers, mostly young, lived in dormitories, ate in shifts in gigantic communal canteens and often worked eighty- to one-hundred-hour weeks. Apple had its Nike moment, becoming the focus of anger about digital sweatshops and worker exploitation, even though Foxconn's factories make products for almost all the major electronics firms, not just Apple.

Long before the controversy, however, Apple helped change the game in manufacturing, forging close relationships with Chinese manufacturers, constantly pushing the limits of what their factories are capable of doing. But the process involved a certain amount of bemusement as to the ways of the Asian companies.

Satzger for one was both impressed by Foxconn and amused at how they did things. "There was a lot of politics," he said. "They would stage things for us. Lots of theatrics. They would bring managers into meetings and yell at them in front of the Apple group."

On one occasion, the Apple group was carefully positioned outside of a conference room with big glass windows. Inside, a Foxconn executive was yelling at his staff. As the Apple group looked on, he brought his fist down on the glass conference table and shattered it.

"He completely set that up," said Satzger.

Foxconn has a mixed reputation, but Satzger said all the engineers—everyone who dealt with Apple at a higher level—were really happy and engaged. However, the assembly plants have been the critical lightning rod, as thousands of workers perform repetitive tasks assembling the products.

The designers spent a lot of time flying out to China to work with Foxconn and other contractors. Satzger would never spend more than five days at a time there. If need be, he'd fly back to California for the weekend, then fly out again the following week. Jony, on the other hand, would sometimes be gone for weeks, and some of the other designers for months. Of course, they were pioneering new materials and new means of manufacture for Apple's extraordinary products.

Over the years, the elevated status of Jony's group would be reflected in their accommodations. According to a former operations engineer, "When all of the teams went to China, the PD [product design] guys and the ID guys, we usually all worked together in the factory. But when we left, the ID guys got picked up in stretch limos and we had to take a taxi," the source said. "All the ID guys stayed in 5-star hotels while most of the PD guys stayed in 3-star hotels. This was different from 10 to 15 years earlier, when in the middle of the build-up to make the iMac, all the designers, including Jony and Danny Coster, lived and ate in the same hotels as the engineers and the rest of the Apple crew."

Jony Ive *(circled)* as a student at Walton High School in Stafford, UK.

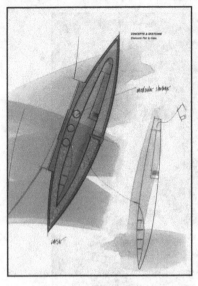

A concept sketch for an electronic pen that wrote in different line widths and patterns.

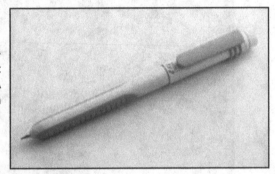

Jony's Zebra TX2 pen had a special mechanism at the top just for its owners to fiddle with. A tactile fiddle factor would be an ongoing motif in Jony's work.

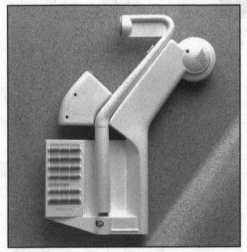

One of Jony's projects at Newcastle reimagined the landline telephone. He called it the Orator.

These are some of Jony's initial sketches for a power drill for UK manufacturer Kango.

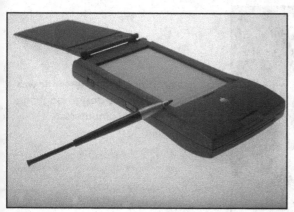

Jony's first major project at Apple was the redesigned Lindy MessagePad 110, which won him a bunch of design awards—but the product never really took off.

Like a lot of prototypes, Jony's Lindy MessagePad was made in clear acrylic to check its thermal heat-dissipation.

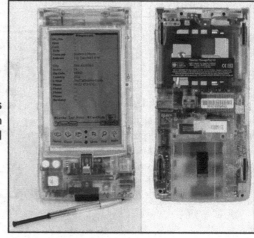

This Baby Mac from frog design is the precursor to the iMac, and a good example of the Snow White design language. Steve Jobs was working on it when he quit/was fired from Apple in 1985.

Frog design's Snow White aesthetic was so influential it set the design language for a generation of computers.

When Jony Ive joined Apple in 1992, the design team was slowly trying to move away from Snow White which had dominated the '80s.

The Domesticated Mac was one of Jony Ive's first speculative designs for Apple. It was an attempt to design a computer for the home, not an office environment.

Another of Jony's early major projects, the Twentieth Anniversary Mac was Apple's first flatscreen computer. It was also designed for the home, not the office, but bungled pricing and marketing doomed it.

The eMate was Apple's first translucent product. Jony felt that translucency made a product less mysterious and more accessible.

Jony Ive (left) with his former boss Jon Rubinstein, head of engineering, with some multicolored iMacs, the first product to bring fashion to computers.

Almost as soon as Steve Jobs returned to Apple in 1997, he formed a deep and productive bond with Jony. The pair shared a fascination with and delight in products and design.

This unusual transparent iBook reveals the complex internal metal frame Jony's design team developed for the iBook, which was a smash success.

The distinctive Luxo Lamp iMac G4 was Jony's second attempt at a flatscreen computer for the home.

For a joke, the IDg team designed the inside of the iMac G4's box to look like male genitals.

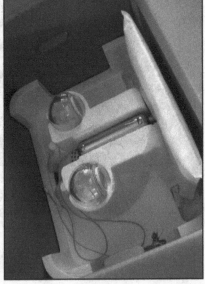

An engineering prototype of the Power Mac G4 Cube. Jony and Jobs hoped it would be Apple's ultimate computer, but it bombed.

An early engineering prototype of the G4 Cube, and an early design prototype.

The guts of the Power Mac Cube. The Cube was an attempt to cram the guts of a desktop computer into a much smaller space.

The Power Mac G5 was the first computer to feature an interior that was designed entirely by Jony Ive's team to be aesthetically pleasing.

This early engineering prototype of the iPhone was built to test a lot of new components, like the custom-made ARM chip.

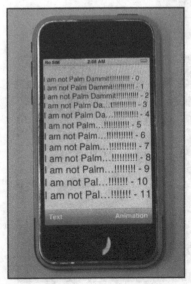

This iPhone prototype made in late 2006—just six months before the iPhone went on sale—featured a plastic screen. At the last minute it was switched to much more durable glass.

One of the hardest design challenges was fixing the gap between the glass screen and the stainless steel bezel. Jony's team kept getting their designer stubble caught in the gap.

The prototype iPhone was tested with an early beta version of iOS, which had little of the fit and finish of the ultimate product.

This iPhone prototype is running a beta version of iOS.

This prototype iPad has two dock connectors: one on the bottom and one on the side.

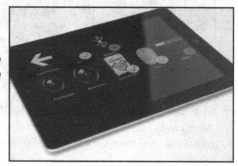

Jony is joined by some of his design team at San Francisco's Apple store, including Daniele De Iuliis *(far left)*, and Danny Coster and Peter Russell-Clarke *(right)*. They are celebrating the first day of iPad sales, which designer Chris Stringer described as a "very special day."

Along with Jony, Chris Stringer *(left)* and Richard Howarth *(right)* are said to be the core members of Apple's industrial design team.

Associated Press/Rex Features

Sir Jonathan Ive after getting his knighthood with his good friend, designer Marc Newson (who received a CBE for services to design).

Jony joins Steve Jobs in a product demonstration shortly before Jobs's untimely death.

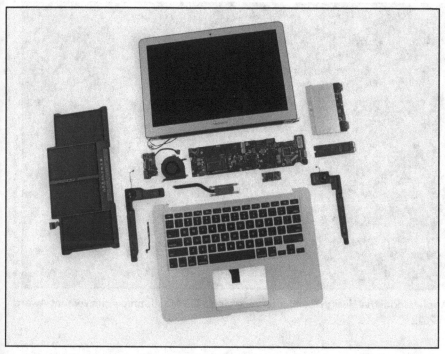

The MacBook Air was one of the first Apple products to utilize a Unibody design, a major breakthough that reduces a complex computer case to a single hunk of finely machined metal.

Apple's industrial design team, after receiving a D&AD Lifetime Achievement Award in 2012.

CHAPTER 10

The iPhone

When we are at these early stages in design . . . often we'll talk about the story for the product— we're talking about perception. We're talking about how you feel about the product, not in a physical sense, but in a perceptual sense.

—JONY IVE

One morning in late 2003, just before the launch of the iPod mini, Jony and his team gathered for a biweekly brainstorming meeting. As usual, the team assembled around the studio's kitchen table. One of the industrial designers, Duncan Kerr, did a show-and-tell. Kerr, who joined Apple's design team in 1999 after having spent a few years working at IDEO, had a lot of engineering experience, and he loved to tinker with new technology.

Kerr had been working with Apple's input engineering group, which was exploring alternative inputs for the Mac, with the hope of doing away with the keyboard and mouse, the mainstay of computing for more than three decades. When Kerr told the group about what he'd learned, his words were greeted by some stunned expressions.

"It was amazing," said Doug Satzger, shaking his head in disbelief. "It was a really amazing brainstorm."

Around the table was the core IDg: Jony, Richard Howarth, Chris Stringer, Eugene Whang, Danny Coster, Danny De Iuliis, Rico Zorkendorfer, Shin Nishibori, Bart Andre, and Satzger.

"I remember Duncan showing us how, with multi-touch, you could do different things with two fingers and with three fingers," recalled Satzger. "He showed us on-screen rotating and zooming—and I was really surprised that we could do that stuff."

That morning was the first time the team had even heard of multi-touch. Today it doesn't seem exceptional, but back then, touch interfaces were pretty primitive. Most touch devices, such as Palm Pilots and Windows tablets, used a pen or stylus. Screens that were sensitive to fingers, not pens, like ATM screens, were restricted to single presses. There was no pinching or zooming, no swiping up and down or left and right.

Kerr explained to his colleagues that the new technology would allow people to use two or three fingers instead of just one, and that it would afford much more sophisticated interfaces than simple single-finger button presses.

Excited by Kerr's explanation of what a sophisticated touch interface could do, the team members started to brainstorm the kinds of hardware they might build with it. The most obvious idea was a touch-screen Mac. Instead of a keyboard and mouse, users could tap on the screen of the computer to control it. One of the designers suggested a touch-screen controller that functioned as an alternative to a keyboard and mouse, a sort of virtual keyboard with soft keys.

As Satzger remembered, "We asked, 'How do we take a tablet, which has been around for a while, and do something more with it?' Touch is one thing, but multi-touch was new. You could swipe to turn a page, as opposed to finding a button on the screen that would allow you turn the page. Instead of trying to find a button to make operations, we could turn a page just like a newspaper. I was really surprised you could do this stuff."

Jony in particular had always had a deep appreciation for the tactile nature of computing; he had put handles on several of his early machines specifically to encourage touching. But here was an opportunity to make the ultimate tactile device. No more keyboard, mouse, pen or even a click wheel—the user would touch the actual interface with his or her fingers. What could be more intimate?

The input engineering team had built a giant experimental system to test multi-touch. It was a big capacitive display about the size of a Ping-Pong table, with a projector suspended above it. The projector shone the Mac's operating system onto the array, which was a mass of wires.

"This is going to change everything," Jony told the design team after he saw it.[1] Jony wanted to show the system to Steve Jobs, but he was afraid his boss would pour cold water on it because it was still raw and unpolished. Jony reasoned that he had to show the work in progress to Jobs in private, with no one else around. "Because Steve is so quick to give an opinion, I didn't show him stuff in front of other people," Jony said. "He might say 'This is shit,' and snuff the idea. I feel that ideas are very fragile, so you have to be tender when they are in development. I realized that if he pissed on this, it would be so sad because I know it was so important."[2]

Jony followed his instincts and showed Jobs the system in private. The gambit worked, and Jobs loved the idea. "This is the future," said Jobs.[3]

With Jobs's seal of approval, Jony directed Imran Chaudhri and Bas Ording, two of Apple's most talented software engineers, to shrink the massive capacitive array into a working tablet prototype. Within a week, they came back with a twelve-inch MacBook display hooked to a big tower Power Mac, which provided the computing power to interpret the finger gestures.

They showed Jony and the designers a demonstration with Google Maps. After bringing up Apple's Cupertino HQ, one of them spread his fingers apart on the screen, zooming in on the campus. The designers were astonished. "We could zoom in and zoom out with touch gestures onto the Apple campus!" said Satzger.

Building a finger-controlled tablet looked like a real possibility. It wouldn't happen overnight and, thanks to market forces, another revolutionary Apple product would emerge from the pipeline first.

Model 035

Multi-touch might have been new to Jony's design team, but it wasn't new in academia. The origins of the technology stretched back to the sixties, when researchers worked out the first crude electronics for touch-based sensors. Systems that could detect multiple touches simultaneously were invented in 1982 at the University of Toronto, and the first workable multi-touch screens appeared in 1984, the same year Steve Jobs launched the Macintosh. The marketplace didn't see multi-touch products until the late nineties. Among the first were a gesture-based input pad for computers and a touch-sensitive keyboard-cum-mouse, from a small Delaware company called FingerWorks.

Early in 2005, Apple quietly acquired FingerWorks and immediately pulled its products from the market. News of the buyout didn't leak for more than a year, when the two FingerWorks founders, Wayne Westerman and John Elias, started filing new touch patents for Apple.

After Chaudhri and Ording's crude mock-up showed that a finger-controlled tablet would work, Jony's industrial design team set about building more finished prototypes. Bart Andre, who also has a mechanical bent, and Danny Coster led the design work. One of the prototypes they created, known internally as "Model 035," formed the basis for a patent filed on March 17, 2004.

Model 035 was a large, white tablet that looked like the lid of one of Apple's white plastic iBooks from the time. Though it lacked a keyboard, it was based on iBook components. The 035 had no home button and a significantly thicker and wider base than would the 2010 iPad. But the two devices share rounded edges and a black bezel surrounding the screen. It ran a modified version of Mac OS X (the mobile version of the software, iOS, was still years away).

While Jony's team worked on several tablet prototypes, Apple's executives were worrying about the iPod. It was flying high: Apple sold two million in 2003, ten million in 2004, and forty million in 2005. But it was becoming clear that the mobile phone would one day supersede the iPod. Most people were carrying around both an iPod and a cell phone. At that stage, cell phones could store a few tunes, but it was becoming clear that, sooner rather than later, someone, perhaps a competitor, would combine the two devices.

In 2005, Apple teamed up with Motorola to release an "iTunes phone" called the ROKR E1. It was a candy-bar-shaped phone that could play music purchased from the iTunes Music Store. Users could load songs through iTunes and play them through an iPod-like music app. But the limitations of the phone doomed it from the start. It could hold just one hundred songs, transferring songs from a computer was slow and the interface was horrible. Jobs could barely conceal his disdain for it.

On the other hand, the Motorola ROKR phone made it apparent to all concerned that Apple needed to make its own phone. Customers wanted the experience of a full iPod on their phones, but, given Jobs's insistence on Apple standards, another company could hardly be trusted to get it right.

Precisely how the project that had produced Model 035 got re-tracked into making the iPhone is a matter of dispute. During an appearance at the 2010 All Things Digital conference, Jobs took credit for having come up with the idea for a touch-screen phone.

"I'll tell you a secret," Jobs told the crowd. "It began with the tablet. I had this idea about having a glass display, a multi-touch display you could type on with your fingers. I asked our people about it. And six months later, they came back with this amazing display. And I gave it to one of our really brilliant UI guys. He got scrolling working and some other things, and I thought, 'My God, we can build a phone with this!' So we put the tablet aside, and we went to work on the iPhone."[4]

Others at Apple at the time have a different recollection of the beginnings of their iPhone pursuit. They say the idea came up during one of the regular executive meetings. "We all hated our phones," recalled Scott Forstall, a software executive. "I think we had these flip phones at the time. And we were asking ourselves, 'Could we use the technology we were doing with touch that we'd been prototyping for this tablet and could we use that same technology to build a phone, something the size that could fit in your pocket, but give it all the same power that we were looking at giving to the tablet?'"[5]

After the meeting, Jobs, Tony Fadell, Jon Rubinstein and Phil Schiller went over to Jony's studio to see a demo of the 035 prototype. They were impressed by Jony's demonstration of the 035, but expressed doubts that the technology would work for a cell phone.

The crucial breakthrough was the creation of a small test app that used only part of the 035 tablet's screen. "We built a small scrolling list," said Forstall. "We wanted it to fit in the pocket, so we built a small corner of it as a list of contacts. And you would sit there and you'd scroll on this list of contacts, you could tap on the contact, it would slide over and show you the contact information, and you could tap on the phone number and it would say 'Calling.' It wasn't calling, but it would say it was calling. And it was just amazing. And we realized that a touch screen that was sized, that could fit into your pocket, would work perfectly as one of these phones."

Years later, Apple attorney Harold McElhinny would describe the immense amount of work the project required. "It required an entirely new hardware system. . . . It required an entirely new user interface and that interface had to become completely intuitive." He also said Apple took a huge leap of faith moving into a new product category. "Think about the risk. They were a successful computer company. They were a successful music company. And they were about to enter a field that was dominated by giants. . . . Apple had absolutely no name in the [phone] field. No credibility."[6]

McElhinny also said he firmly believes that had the project gone wrong, it could have destroyed the company. To mitigate the risk, Apple's executives hedged their bets. They would develop two phones in parallel and pit them against each other. The secret phone project was code-named Purple, shortened to just "P." One phone project, based on the iPod nano, got the code name P1; the other phone, led by Jony, was a brand new multi-touch device based on the 035 tablet, code-named P2.

The P1 project was led by Fadell; his group had the idea to somehow graft a phone onto a current iPod. "It was actually a natural progression of taking the iPod, which we already had, and morphing it into something else," said the former executive.

Matt Rogers, a hotshot young iPod engineer who worked for Fadell, was given the job of creating the software for the device. As an intern, Rogers had previously impressed Fadell by rewriting some complex testing software for the iPod. As usual, the research was a big secret. "Nobody in the company knew we were working on a phone," said Rogers.[7] It was also a lot of extra work. At the time, the iPod team was also working on a new iPod nano, a new iPod classic and a Shuffle.

After six months of effort, Fadell's team produced a prototype iPod-plus-phone that worked, more or less. The iPod's click wheel was used as a dialer, selecting numbers one at a time like an old rotary phone. It could make and receive calls. Scrolling through an address book and selecting a contact to call was—unsurprisingly—its best feature. Apple filed a couple of patents from their experimentation. One of them suggested that the iPod-plus-phone could create text messages with a predictive text system. Jobs, Forstall, Ording and Chaudhri, among others, were named as inventors.

But the P1 had too many limitations. Just dialing a number was a pain, and the device was too limited. It couldn't surf the Net; it couldn't run apps. Fadell said later that the iPod-plus-phone was a "heated topic" of discussion at Apple. The biggest problem was that it had forced the team into a design corner. Using the existing device limited their design options in a way that was not optimal to the task. "[The P1] had a little screen and this hardware wheel and we were stuck with that . . . but sometimes you have to try things in order to throw it away."[8]

After six months of work on the iPod-plus-phone P1, Jobs killed the project. "Honestly, we can do better, guys," he told the team. Fadell was loath to admit defeat. "The multi-touch approach was riskier because no one had tried it and because they weren't sure they could fit all the necessary hardware into it," he said. And Fadell had been skeptical of touch screens from the start, based on experience of devices like Palm Pilots, which were clunky and awkward.

"We all know this is the one we want to do" Jobs said, referring to the P2. "So let's make it work."[9]

Two years later, during the iPhone introduction at Macworld, Jobs jokingly flashed an image of an iPod with a rotary dial pad on its screen. This was how not to build a new phone, Jobs said, as the audience laughed. Few knew the company might well have produced just such a phone.

A New Team Takes Charge

After the decision to move forward with the P2, Jony was put in charge of industrial design, Fadell of engineering and Forstall, previously responsible for Mac OS X, was given the job of adapting the computer operating system into a brand-new operating system for the phone.

Jony's design team worked on the iPhone without ever seeing the operating system. They initially worked with a blank screen and later, a picture of the interface with cryptic mock icons. Likewise, the software engineers never got to see the prototype hardware. "I still don't know what the lightning icon means," one of the designers later remarked, referring to one of the icons on the fake iOS screen.

Jony himself wasn't left in the dark: He was kept up to speed on the latest developments in Forstall's new operating system, and was constantly talking to Jobs and other executives. He'd give feedback and direction to the design team. Within the design studio, designer Richard Howarth was designated the design lead of the P2.

At the beginning, few of those involved were confident they would be able to develop a phone. "It was fundamental R&D in all directions," said a former executive. It meant ramping up probably the most difficult project in the company's history and, all the while, continuing to develop products like the MacBook and the iPod line. Important staffers were moved off their current projects, delaying some products and canceling others.

There were potentially dire consequences for the company if the project did not succeed. "Had it not succeeded, not only would we have had the detriment of the lack of those products shipping, we wouldn't have had something else to fill in at the same time," Forstall explained.

Jobs told the executives they could recruit anyone they wanted within the company to work on the project, but they absolutely could not go outside. "That was quite a challenge," Forstall recalled. "The way I did it is I would find people who were true superstars at the company, just amazing engineers, and I would bring them to my office and I would sit them down and I would say, 'You are a superstar in your current role. Your manager loves you. You're going to be incredibly successful at Apple if you just stay in your current role and keep on doing what you want to do. I have another offer for you, another option. We're starting a new project. It's so secret, I can't even tell you what that new project is' . . . and amazingly, some tremendously talented people accepted that challenge and that's how I put together the iPhone team."[10]

Forstall commandeered an entire floor in one of the buildings at Apple HQ and had it locked down. "We put doors with badge readers, there were cameras, I think, to get to some of our labs, you had to badge in four times to get there," he said. It was nicknamed the "Purple Dorm."

"People were there all the time," Forstall said. "They were there at night. They were there on weekends. You know, it smelled something like pizza.

"On the front door of the purple dorm, we put a sign up that said 'Fight Club' because the first rule of *Fight Club* in the movie is you don't talk about Fight Club, and the first rule about the purple project is you do not talk about that outside of those doors."

Over in IDg, Jony began, as usual, with the iPhone's story. As he later explained, it was all about how the user would *feel* about the device. "When we are at these early stages in design, when we're trying to establish some of the primary goals—often we'll talk about the story for the product—we're talking about perception. We're talking about how you feel about the product, not in a physical sense, but in a perceptual sense."

Jony believed the iPhone would be all about the screen. In their earliest discussions, the designers agreed that nothing should detract from the screen, which Jony likened to an "infinity pool," those high-end swimming pools with an invisible edge.

"What that did was make it very clear in our minds that the display was important, and we wanted to develop a product that featured and deferred to the display," he said. "Some of our early discussions about the iPhone centered on this idea of . . . this infinity pool, this pond, where the display would sort of magically appear."[11] The team made a point in exploring design ideas to avoid any approaches that would diminish the importance of the display.

Jony said they wanted the display to be "magical" and "surprising." These were his high-end goals for any eventual design. "In the earliest stages of this design, this seemed—this was very new, and it felt there was real opportunity to develop a design story based on those sorts of preoccupations," he later explained.[12]

Late in the fall of 2004, Jony's design team began work on two distinct design directions. One, called Extrudo, was led by Chris Stringer, and it resembled the iPod mini. It was made from a flattened tube of extruded aluminum and could be anodized in different colors. Apple already had big production lines making and anodizing iPod cases in huge numbers. That was one advantage of that direction, along with the fact that Jony and team loved what could be done with extrusion.

The other design, called Sandwich, was led by Richard Howarth. Made mostly of plastic, with a plastic screen, the Sandwich design was rectangular with evenly rounded corners. It had a metal band running around the midpoint of its body, a centered display on the front face, a menu button centered below the screen and a speaker slot centered above the screen.

Jony and his team preferred the Extrudo look and gave it the most attention. They tried cases that were extruded along the x-axis, and some along the y-axis. But problems surfaced immediately. Extrudo's hard edges hurt the designers' faces when they put it up to their ears. Jobs especially hated this.

To make the hard edges softer, plastic end caps were added, which also helped with the radio antennas. The iPhone would have three radios: Wi-Fi, Bluetooth and a cell radio. But radio waves won't pass through a metal shell, so the plastic endcaps became essential.

The team struggled to solve Extrudo's problems, but engineering tests made it clear this particular design direction wouldn't work unless the plastic endcaps for the radios got bigger. But bigger caps would ruin the clean Extrudo look. "We made books and books filled with pages of designs trying to figure out how not to break up the design because of the antenna, how not to make the earpiece too hard and sharp, and so on," said Satzger. "But it seemed like all the solutions that added comfort detracted from the overall design."

The Extrudo design had another problem that nagged at Jobs: The metal bezel detracted from the screen. The design didn't "defer" to the screen, which had been one of Jony's original goals. Jony later recalled his flush of embarrassment when Jobs pointed it out.

Apple killed Extrudo; the team was left with Sandwich.

The Sandwich design did have several advantages over Extrudo, one of which was that the rounded edges didn't hurt the designers' ears. But the engineering prototypes came back big and chunky, and Jony's team struggled to slim it down. They were trying to cram in a lot of technology, much of which hadn't yet been miniaturized enough for a device as complex as the phone everyone envisioned.

The Jony Phone

By February 2006, several redesigns had come and gone. Jony was so dissatisfied with the progression that, during one of the brainstorms, he asked designer Shin Nishibori to make an exploratory version of the phone with Sony-style design cues. Later, he would contend that his request was not to copy Sony specifically, but to inject some fresh, "fun" ideas into the process.

Shin Nishibori had been a well-known young designer in Japan for years before coming to work at Apple. Traces of a Sony/Japanese influence have appeared in Nishibori's work on Apple products since 2001, and Steve Jobs, Jony and other Apple designers had often expressed admiration for Japan's minimalist aesthetic.

In February and March 2006, Nishibori designed and built several phones that borrowed elements seen in Sony products of the time, including a jog wheel, which was a control-wheel-cum-switch that was used on Sony's CLIÉ personal digital assistants. Nishibori even put the Sony logo on the backs—except for one that he jokingly labeled as Jony.

Years later, at the *Apple v. Samsung* trial, one of Nishibori's mock Sony phones would be presented as evidence that Jony's design team hadn't developed the iPhone on their own, as they contended, but instead copied other companies' designs. But Apple successfully argued that the Sony/Jony design was merely Sony-style decoration on a device they'd already designed.[13] As Apple's attorneys pointed out, Nishibori's designs are asymmetrical and none of the Sony-style buttons and switches were adopted in the released iPhone.

In early March 2006, Richard Howarth expressed his frustration with the state of the development of P2. In comparing P2 to Nishibori's Sony-style design, Howarth complained about its size, and how Nishibori had managed to achieve a slimmer profile. "Looking at what Shin's doing with the Sony-style chappy he's able to achieve a much-smaller looking product with a much nicer shape to have next to your ear and in your pocket," Howarth wrote in an e-mail to Jony.

"I'm also worried that if we start cutting volume buttons on the side, then it removes some of the purity of the extrusion idea and seems like the wrong shape for the job. We can only add so much to it before it becomes a style/a shape, rather than the most efficient construction method and that would be bad."[14]

Jony's team had also tried a curved design, which at one point looked like a promising direction. By adding a curve, more technology could be packed into the middle bulge. It's a trick Apple has used in many of its newer products, from the iPad to the iMac.

From the beginning, Satzger remembered, the team had a "strong interest" in a design that used two pieces of shaped glass. One of the prototypes they built had a split screen. Above was the screen, below a software-driven touch pad that changed depending on function. Sometimes it was a dial pad, at other times a keyboard. But the problem of making glass convex proved too difficult.

Although Howarth was still using the Extrudo design as a comparison piece late in development, engineering tests seemed to affirm that the Sandwich-style direction would prevail. Then engineering prototypes of the Sandwich design came back with a bad report: They were too big and fat. With the realization that all the technology just couldn't be squeezed into a pleasing shape, the decision was made to kill the Sandwich design too.

"We didn't know enough about antennas, we didn't know enough about acoustics, we didn't know enough about packing everything in," said one former executive. "It worked, but it just wasn't attractive."

Faced with a dead end, Jony's team reversed course. They turned to an old model they'd made early in the process but had discarded in favor of the Sandwich and Extrudo. The discarded model looked very similar to the phone that would actually ship, with an edge-to-edge screen interrupted only by the single home button. Its gently curved back snapped seamlessly onto the screen, like the original iPod. Most importantly, it maintained Jony's infinity-pool illusion. When the phone was off, it appeared to be a single, unbroken, inky-black faceplate; when switched on, the screen magically appeared from within.

It was a voila! moment. "We found something that we'd overlooked," said Stringer, "something that we, once adding detail to it and really spending some time with it, decided was the absolute best choice for us at that time." He remembered the ease with which the final, unornamented design for iPhone was chosen. "It was the most beautiful of our designs," he explained. The front face bore neither the company logo nor the name of the product. "We also knew from our experience with iPod," Stringer explained, "if you make a startlingly beautiful and original design, you don't need to. It stands for itself. It becomes a cultural icon."

In the fullness of time, Jony resurrected the Sandwich for the iPhone 4, another example of the team revisiting earlier designs and spying qualities that they'd previously overlooked. The main structural element of the iPhone 4—the steel frame between the two plates of glass—would double as the phone's antennas. Unfortunately, that proved problematic because, if the user's hand made a contact between a gap, effectively shorting the two antennas, the phone would drop calls. Reportedly, Apple could have easily avoided the issue if the antennas had been clear coated, but Jony didn't want to ruin the integrity of the metal.

Beyond matters of form, the team focused on the function of the multi-touch. Most touch devices at the time used resistive touch screens, based on two thin sheets of conductive material separated by a thin gap of air. When the screen is pressed, the two layers make contact, registering the touch. Resistive screens were typically made of plastic, and were common in pen-based devices like Palm Pilots and Apple's Newton.

Jony's design team tried using a resistive screen for the iPhone, but were unsatisfied with the results. Pressing on the screen distorted the picture, and the screen tired the fingers because the user had to press pretty hard. It just didn't live up to the promise of the name ("touch screen"), which the designers thought should convey the illusion that the user was literally touching the content.

Moving on from the resistive screen, the hardware team set out to build screens based on capacitive touch, registering changes in electrical charges (or capacitance) across its surface. Human skin is electrically conductive, and a capacitive touch screen uses that characteristic to detect even the lightest touch. Apple had been using capacitive touch technology for several years with the iPod scroll wheel, laptop track pads and the Power Mac Cube, which had a capacitive on-off button. But the technology hadn't been applied to transparent screens.

One problem was that there was no supply chain for capacitive screens. No one was producing them on an industrial scale at the time—but Apple found a small company in Taiwan called TPK that was producing them for point-of-sale displays using an innovative but limited-run technique. Jobs made a handshake deal with the company, promising that Apple would buy every screen the factory could produce. Based on this agreement, TPK invested $100 million to rapidly ramp up their manufacturing capabilities. They ended up supplying about 80 percent of the screens for the first iPhone, growing rapidly to a $3 billion business by 2013.

From Plastic to Glass

While Apple's operations group was working out how to manufacture the iPhone, Jony's design team was having doubts about their original choice of material for the screen.

Jony and his team planned to use plastic, mostly because it was shatterproof. Although all of the iPhone prototypes had plastic screens, the designers were never happy with it.

"The original plastic face had this weird flexibility to it," said Satzger. "It was a matte finish. If it goes to glossy plastic, you see this waviness to it, which makes it look really crappy." Jony instructed the team to try textured plastic, but that didn't work either. In a gutsy next move, they decided to try glass, despite the facts that glass breaks easily and no one had made a consumer electronic device with such a big piece of glass.

The story of the shift to glass is variously remembered: Although Jony's group was already investigating glass, Jobs is credited by some as initiating the move to glass. As the story goes, he'd been using an iPhone prototype, which he kept in his pocket with his keys. He was reportedly furious that they scratched the screen.

"I won't sell a product that gets scratched," Jobs said later. "I want a glass screen, and I want it perfect in six weeks."[15]

A deeper version has it that the development took more like six months in advance of launch, not six weeks. Apple's operations group was charged with finding the strongest glass available. The search led them to Corning Incorporated, a glass manufacturer headquartered in upstate New York.

In 1960, Corning had created a nearly unbreakable reinforced glass they called "muscled glass" or Chemcor. The key to its manufacture was an innovative chemical process in which glass is dipped into a hot bath of potassium salt. Smaller sodium atoms leave the glass, and are replaced by bigger potassium atoms from the salt. When the glass cools, the bigger potassium atoms are packed in so tightly they give the glass exceptional damage resistance. The glass can withstand pressures of 100,000 pounds per square inch (normal glass handles about 7,000). Chemcor's market future seemed bright, but it never took off. Other than its use in some airplanes and American Motors Javelin cars, the material sold poorly and Corning discontinued it in 1971.[16]

When Apple's operations group came calling in 2006, they found Corning had been thinking about bringing back the old superglass for a couple of years. They'd seen Motorola use glass for its RAZR V3 phone and begun to explore ways to make Chemcor thin enough to be suitable for cell phones.

Jony's ID group came up with spec: The glass needed to be 1.3 mm to fit into the iPhone design. Jobs told Corning's CEO, Wendell Weeks, they had six weeks to create as much of it as they could. Weeks replied they didn't have the capacity and actually, Chemcor had never been created in this way or in such volume. "None of our plants make the glass now," he told Jobs. But Jobs cajoled the CEO. "Get your mind around it," he said. "You can do it."[17]

Almost overnight, the company completely remade its manufacturing processes, based in Kentucky, changing several of its LCD-making plants to muscled glass, by then renamed Gorilla Glass. In May 2007, Corning was making thousands of yards of Gorilla Glass.

Corning's glass, in combination with the aluminum back, marked another change in Jony's design language. It was a striking, almost shocking, minimalism in hard metal and glass.

To hold the glass screen in place, Jony's team came up with a shiny stainless steel bezel, which doubled as a structural element. The bezel would give the iPhone strength, but it also needed to look good.

Jony's team worried that the glass would smash if the phone was dropped. "We were putting glass in close proximity to hardened steel," said Satzger, who pointed out that, "if you drop [the phone], you don't have to worry about the ground hitting the glass. You have to worry about the band of steel surrounding the glass hitting the glass."

The solution was a thin rubber gasket between the glass screen and the stainless steel bezel. But the gasket created a gap that, at least at first, the designers hated for a very personal reason. "Because many of us in the ID team rarely shaved and had beard stubble, it used to yank our facial hair when we held the device up to our faces," said Satzger, laughing at the memory. The team played around with gap size until they got it right.

"We designed several increments of gap size between the metal and glass until we got one that didn't yank our face hair."

Countdown to Macworld

One morning in the fall of 2006, Jobs gathered the iPhone leaders in Apple's boardroom to talk about the state of iPhone development. Fred Vogelstein of *Wired* described the scene as a horror show of bad news.

"It was clear that the prototype was still a disaster. It wasn't just buggy, it flat-out didn't work. The phone dropped calls constantly, the battery stopped charging before it was full, data and applications routinely became corrupted and unusable. The list of problems seemed endless. At the end of the demo, Jobs fixed the dozen or so people in the room with a level stare and said, 'We don't have a product yet.'"[18]

The fact that Jobs seemed calm—instead of bringing his usual fire and brimstone—spooked everyone in the room. Vogelstein said one executive described the moment as "one of the few times at Apple when I got a chill."

Because the iPhone announcement was going to be the main event at Macworld in a few weeks, any sort of postponement would have been disastrous. "For those working on the iPhone, the next three months would be the most stressful of their careers," Vogelstein wrote. "Screaming matches broke out routinely in the hallways. Engineers, frazzled from all-night coding sessions, quit, only to rejoin days later after catching up on their sleep. A product manager slammed the door to her office so hard that the handle bent and locked her in; it took colleagues more than an hour and some well-placed whacks with an aluminum bat to free her."

The problem was that everything was new and nothing worked. The touch screen was new, so were the accelerometers. The proximity sensor, which turned off the screen when the user held the phone up to their face, developed a problem in a late prototype: It worked for most people, but it didn't work if the user had long dark hair, which confused the sensor.

"We nearly shelved the phone because we thought there were fundamental problems that we can't solve," Jony told a business conference in London. "You have to detect all sorts of ear-shapes and chin shapes, skin color and hairdo . . . that was one of just many examples where we really thought, perhaps this isn't going to work."[19]

But just weeks before Macworld, Jony's team had a prototype that worked well enough to show AT&T. In December 2006, Jobs traveled to Las Vegas to show it to the wireless carrier's CEO, Stan Sigman, who was "uncharacteristically effusive," calling the iPhone "the best device I have ever seen."[20]

The arrival of the iPhone at Macworld was the culmination of more than two and half years of intense hardship, learning and dedication to bring it to market. As one Apple executive summed it up, "Everything was a struggle. Every. Single. Thing was a struggle for the entire two-and-a-half years."

When launch day came in mid-summer 2007, Jony joined the whole design team at the flagship Apple retail store in San Francisco. "We were excited," said Stringer. "We had something new. There was an incredible buzz. . . . And there was an enormous crowd outside. We wanted to feel that enthusiasm and see people, see their eyes when they get these new products, the first people to get them. When the doors opened, there was mayhem. It was like a carnival."[21]

Stringer was overwhelmed with emotion. "We were obviously very, very proud. We'd worked really hard. It was—there was an enormous number of people that put in personal sacrifice and it was paying off in spades. It was a beautiful day."

*   *   *

The iPod had been regarded by a lot of pundits as Apple getting lucky, a fluke, a one-shot. When Apple entered the cutthroat cell phone market, it was predicted the iPhone would flop. Microsoft's Steve Ballmer famously said it would never get any market share. But the iPhone was a hit from the start, and Apple used its old playbook of rapidly adding features and models.

Apple released the iPhone in mid-2007. By the end of the year, 3.7 million iPhones had been sold. By the first quarter of 2008, the sales volume of iPhones exceeded sales of Apple's entire Mac line. And by the end of 2008, the company was selling three times as many iPhones per quarter as it was selling Macs. Revenue and profits were through the roof.

When Jobs unveiled the iPhone at Macworld in January 2007, he invited his old friend Alan Kay to the launch. Jobs and Kay knew each other from Xerox PARC, and later Kay had been appointed an Apple fellow, a kind of elder statesman, and worked for a decade inside Apple's Advanced Technology Group in the late nineties. Kay is famous for prophesizing the "Dynabook," a tablet computer that would provide a window into all the world's knowledge—back in 1968.

On iPhone launch day, Jobs turned to Kay and casually asked, "What do you think, Alan? Is it good enough to criticize?" The question was a reference to a comment made by Kay almost twenty-five years earlier, when he had deemed the original Macintosh "the first computer worth criticizing." Kay considered Jobs's question for a moment and then held up his moleskin notebook. "Make the screen at least five inches by eight inches and you will rule the world,'" he said.[22]

The world would not have to wait very long for the iPad.

CHAPTER 11

The iPad

I can't think of a product that has defined an entire category and then has been completely redesigned in such a short period of time. It is really defined by the display. There are just no distractions.   —JONY IVE

While Jony's group was secretly working on the iPad, Steve Jobs was telling the public and press that Apple had no intention of releasing a tablet. "Tablets appeal to rich guys with plenty of other PCs and devices already," he said publicly. But Jobs was dissembling. "Steve never lost his desire to do a tablet," said Phil Schiller.[1] In fact, while Jony's design team was developing the iPhone, they were also actively working on tablets. Jobs was just waiting for the right time to bring a tablet to market.

One incentive to move forward was the appearance of netbooks, a category of small, inexpensive, low-powered laptops that launched in 2007. They quickly started to eat into laptop sales and, by 2009, netbooks accounted for 20 percent of the laptop market. But Apple never seriously considered making one. "Netbooks aren't better than anything," Steve Jobs said at the time. "They're just cheap laptops."[2] Nonetheless, the subject came up several times in executive meetings.

During one such high-level executive meeting in 2008, Jony proposed that the tablets in his lab could be Apple's answer to the netbook. Jony suggested that a tablet was basically an inexpensive laptop without the keyboard. The idea appealed to Jobs, and Jony was given the go-ahead to transform the prototypes into a real product.

Crucially, mobile technology had advanced significantly in just a few years since the iPhone had been launched. By then, the 035 tablet prototype from 2004 seemed big and unwieldy. But thanks to new screens and batteries, everyone understood that a tablet could be much lighter and slimmer. One of the major reasons the iPad hadn't been green-lighted sooner was that the components like the screen and battery weren't ready. "The technology was not there yet," said a former Apple executive.

Jony began by ordering twenty models made in varying sizes and screen-aspect ratios. They were laid out on one of the studio's project tables for Jony and Jobs to play with. "That's how we nailed what the screen size was," Jony has said.[3] They had done the same thing earlier in finding the right size for the Mac mini and other products.

"Steve and Jony liked to do that with almost all products," said a former engineer in the operations group. "They started off making a bunch of 'appearance' models and they'd make them in all sorts of sizes to find what they want."

But, as often happens, recollections seem to vary. According to an executive at Apple at the time, the screen size was also strongly influenced by a simpler piece of equipment: a standard piece of paper. "The size of the tablet was that of a sheet of paper," he explained. "It was conceived as a legal note tablet, and we thought that was the right size. It was targeted at education and schools and e-reading." Hardware was still another factor, as the guts of the iPad would be based not on the iBook but the iPod touch. Early on, the iPad was understood to be, in effect, a scaled up touch-screen iPod.

Jony's ultimate goal was to make a device that needed no explanation and was fully intuitive. It was to be a "breathtakingly simple, beautiful device, something that you really want, and something that's very easily understandable," Stringer said. "You pick it up, you use it, something that . . . needs no explanation."

That said, producing the "breathtakingly simple" can require an immense investment of time and creative energy.

Making the Machine

Jony's design team explored two different design directions for the iPad, directly akin to the twin design directions they pursued with the iPhone.

Based on the Extrudo design, the first approach built upon a case that resembled the extruded aluminum iPod mini. It was just bigger and flatter. The design lead on this version was Chris Stringer, who also worked on the Extrudo iPhone. As with the phone designs, Stringer's Extrudo iPad was made of a single piece of extruded, milled aluminum. It, too, had plastic caps for the Wi-Fi and cell phone radios. In this case, though, sharp edges weren't much of a concern; no one was going press a tablet up to his or her face.

Jony's IDg team experimented with some "picture frame" models, larger than some of the iPad prototypes, which had kickstands to prop them up. (Kickstands would also feature prominently in competing tablets from Microsoft and other manufacturers in the future.) Jony's team didn't pursue the idea, although adding a kickstand would appear later in the iPad 2's magnetic cover, which could be folded back into a stand.

The designers found Stringer's Extrudo iPad suffered the same limitation as the Extrudo iPhone: The bezel detracted from the screen. As Jony put it, "How do we get out of the way so there aren't a ton of features and buttons that distract from the display?"[4] Again, Jony wanted the infinity-pool illusion because he understood the screen was all-important and that nothing should detract from it.

Meanwhile, Richard Howarth brought his experience with the Sandwich iPhone models to his prototypes, making several versions of Sandwich-style iPads. The early Sandwich iPad models resembled more svelte versions of the 035 prototype. Made of shiny white plastic with a boxy shape, they are clearly in the same design family as Apple's plastic MacBooks, released early in 2006—which makes sense given that they were designed largely by Howarth. Like the plastic MacBook, the device at that point remained fairly big and chunky. Still, Jony's team was clearly homing in on how to present the screen, and the bezel was plain and unobtrusive.

As the design progressed, the new models got thinner, the edges sharper. Some had aluminum backs, but Jony's team seemed to be veering in the direction of the Sandwich. Yet something bothered Jobs: Somehow the iPad wasn't quite casual enough.

Jony spotted the problem. The iPad needed a cue, some sign that it was friendly and could be picked up easily with just one hand. As usual, Jony wanted to invite users to touch the device, pick it up and hold it and have a tactile experience.

The logical next step seemed to be adding handles, and Jony's team experimented with them in an attempt to ease picking up the iPad. One of the later prototypes featured a pair of large plastic handles, making it look like a particularly inelegant TV dinner tray. When they realized the handle approach clearly wasn't working, Jony's team started exploring a tapered back that swept away underneath the screen, opening a gap for fingers to slide underneath.

As Jony's team homed in on the iPad design, they were also completing work on the second-generation iPhone. Marketed as the iPhone 3G, to highlight its compatibility with new 3G cell phone networks, the 2008 follow-up dispensed with the original's aluminum back plate in favor of a hard, polycarbonate plastic. Not surprisingly, then, the two simultaneous development projects shared numerous elements, as the iPad would also get a polycarbonate back, colored black or white, with a stainless steel bezel to marry the back plate to the screen.

Just as they agreed upon a design, however, production problems forced Jony to change it.

The plastic back of the iPhone 3G looks simple, but was extremely hard to manufacture. Jony and the team wanted to use a similar shell for the iPad (comprising a strong blend of polycarbonate and acrylonitrile butadiene styrene), but it proved to be more difficult to manufacture at the larger iPad size, as the larger shell would shrink and warp when it came out of the mold. To stop it from shrinking at the edges, the shell was molded larger than it needed to be and machined down to size.

Even after molding, the shell still had to be polished to remove the part lines, then painted and machined again to prevent the paint shrinking around openings. The manufacturing process gained additional steps, with the openings painted over, then machined out before the installation of the buttons, the speaker grilles and the Apple logo on the back. The use of the plastic had made the entire process problematic. "You have to set those machining processes in the right order because if you machine before you paint, the chemistry of the paint relaxes the surface tension of the plastic and then the sink goes into other areas that you already machined," Satzger said. "It's just easier to do it with aluminum than with plastic."

Jony's team went back to the drawing board and designed an aluminum back. They were comfortable with the material; they already had the process and the production lines down. The new aluminum back wasn't as tapered as Jony would have liked. To give the iPad stiffness, the designers had to add a thin sidewall that gave it strength but made it thicker and bulkier than the planned plastic version.

When they were done, however, Jony's team was excited by the stark minimalism of the device. "We had tried so many things," remembered Chris Stringer. "But at the end of the day, we realized it needed to be its own self. We can't copy ourselves. We wanted a unique form . . . a very anonymous object, not playing along with the lines of consumer electronics at all."[5]

The iPad they produced didn't feel like anything else. As Stringer put it, "It felt like a new object."

iPad Day

On January 27, 2010, Steve Jobs went public with Apple's newest game changer. He announced the iPad at the Yerba Buena Center for the Arts in San Francisco, positioning it as a device that exists between an iPhone and a laptop, a highly portable, handheld slate with a touch-screen interface. He distinguished it from netbooks, describing the iPad as a device more "intimate than a laptop," conveying the sense that the iPad was at the intersection of both technology and art.

The iPad went to market in April. In less than a month, Apple sold one million iPads in half the time it took the iPhone to reach that same mark. By June 2011, just over a year after its release, twenty-five million had been sold. By most measures it became the most successful consumer product launch in history. In 2011, shipments of iPads rapidly overtook those of netbooks, sixty-three million versus fewer than thirty million, according to research firm Canalys.[6]

At Apple HQ, some of the faces who initiated this growth were also changing. In November 2008, Tony Fadell had stepped down as senior vice president of the iPod division, the job he took over from Rubinstein. According to an Apple press release, Fadell and his wife, Danielle

Lambert, who was vice president of the company's human resources department, were "reducing their roles" to "devote more time to their young family."[7] But two former Apple employees say Fadell was another victim of Jobs's close relationship with Jony.

"Tony got canned," said one source. "He was paid off with his salary for a number of years plus so many millions to leave. Tony was canned because he was battling with Jony. He went to Steve so many times bitching about Jony, but Steve had such a tremendous amount of respect for Jony and their relationship that he sided with Jony, not Tony."[8]

iPad Evolution

Less than a year after the iPad's initial launch, in March 2011, Apple surprised the world by announcing a sequel. The new version would be not only a big upgrade in terms of the hardware capability but a whole-sale design turnover.

The iPad 2 was thinner and lighter than the original. It gained key new features like front and back cameras, as well as thoughtful touches like a magnetic cover that turned the iPad off and on. The design marked a big advance in manufacturing (with the unibody process), which allowed Jony to fashion the deeply beveled back he originally wanted, but in metal, using Apple's new unibody manufacturing process. "By reducing what were essentially three surfaces to two, we got rid of the structural wall around the perimeter of the product and eliminated the edge. It's not only more comfortable to hold, but with the breakthrough we made through unibody engineering, it's rigid, sturdy and even more precise."[9]

Once again, Jony was extremely proud of his group's efforts. "I can't think of a product that has defined an entire category and then has been completely redesigned in such a short period of time. It is really defined by the display. There are just no distractions."[10]

In March 2012, Apple followed up with the third-generation iPad, which added a high-density retina display, a faster chip and better cameras. In October of the same year, the fourth-generation iPad was launched with a much faster processor and cell connection, as well as a tiny lightning connector to replace the original thirty-pin connector, which was long in the tooth and had become a legacy technology.

In their constant iterations, Apple was beating the "fast followers" at their own game. Fast followers take a winning product, make it cheaper and get it on the market very quickly. Sometimes the products are cheap knockoffs, but often they are good-enough rivals, as myriad Android phones contest. But by upgrading the iPad quickly and aggressively, and making it significantly better with each version, Apple was staying ahead of their competitors.

The fourth-generation iPad was joined by the iPad mini in 2012, which shrank the screen to just under eight inches, and was enthusiastically snapped up by users. "The Mini gives you all the iPad goodness in a more manageable size, and it's awesome," wrote David Pogue, an influential tech reviewer with the *New York Times*. "You could argue that the iPad Mini is what the iPad always wanted to be."[11] In the first quarter of 2013, the iPad mini accounted for about 60 percent of all iPad sales.[12]

From the initial launch and almost overnight, iPads appeared at cafés and on cross-country flights. Apple executives had predicted several times that the iPad would one day replace the PC, but that switch started happening quicker than anyone expected. In the first year, Apple sold nearly fifteen million iPads. By the fourth quarter of 2011, Apple sold more iPads in just three months than any of its rivals sold PCs. By 2015, tablets (most of them iPads) will have more market share than the entire traditional PC market, according to estimates by the market research firm Interactive Data Corporation. The post-PC era, led by Jony and Apple, is upon us.

CHAPTER 12

Unibody Everywhere

From a design and engineering point of view, Apple is at the absolute pinnacle of creating products that are as close to flawless as can be done.   —DENNIS BOYLE, COFOUNDER, IDEO

In 2008, Jony took the stage at an Apple event to talk about something special: Apple's new "unibody" manufacturing process. His very appearance was a clear sign from the company of the importance of this design breakthrough.

Jony began by talking about the old MacBook Pro, which was one of the lightest and strongest laptops on the market at the time. Its robust strength resulted from a complex structure of internal frames and strengthening plates screwed and welded together. As Jony spoke, a series of slides played behind him, showing the multiple parts layered, bonded and finally mated with a plastic gasket that ran around the middle.

"For years," Jony told the audience, "we have been looking for a better way to make a notebook." He paused and smiled before continuing. "And we think we found it."[1]

Jony went on to explain the manufacture of the MacBook Air, Apple's new razor-thin laptop. Instead of taking multiple sheets of metal and layering them, the new process began with a thick block of metal and, in a reversal of the old process, produced a frame by removing material rather than by adding it. Multiple parts were replaced by just one— hence the name unibody.

Jony's slides illustrated the various stages. Pronouncing aluminum the British way (aloo-min-ium), he said, while smiling: "One of the fantastic things about aluminum is how recyclable it is. So at each of these distinct stages, we are continually collecting the material, and cleaning it and then recycling it."[2]

Although he revealed none of the most essential secrets of the process, he clearly reveled in its fantastic, robot-controlled production. "We started with a solid slab of aluminum, high grade aluminum, that weighed over 2.5 lbs and we end with this remarkably precise part that now only weighs a quarter of a pound. And it's not only incredibly light, it's very, very strong."

The process represented the realization of something else close to Jony's design soul. By melding great design with state-of-the-art manufacturing technology, Apple had produced something remarkable and new. "That one part," continued Jony, "just that single part, forms the structure for the MacBook Air. It really is this highly precise aluminum unibody enclosure that made this product possible."

At that moment, the MacBook Air was the only Apple machine made with the unibody process. However, Apple was about to move almost all of its major products, including the Mac, iPhone and iPad, to unibody.

The change would be a watershed, albeit one that in the excitement surrounding the launch of various products, went largely unacknowledged by the public.

Making the Change

Several months earlier, Jony had laid out all of the parts of a dismantled MacBook Pro on top of one of the big display tables in the design studio at Apple. On the table next to it he had spread out the parts of one of the new unibody MacBooks.

Arranged neatly, the parts of the old MacBook Pro took up almost the entire tabletop. In contrast, the many fewer parts of the unibody machine made for a striking comparison. Jony called his designers over to appreciate the difference.

As part of his characteristic drive to reduce and simplify, Jony wanted to reduce the number of parts and therefore the number of part-to-part joints. Previously, when IDg had done a similar dismantling of an original iPhone, the team counted nearly thirty interfaces where parts meet. After the iPhone underwent a unibody makeover, the number of interfaces shrank to just five.

Jony and his team had initiated the process that led to the unibody much earlier. They'd first explored machining—a manufacturing process that removes raw material to make a part, that may involve drilling, turning, boring and so on—in 2001 with the Power Mac G4 Quicksilver and slowly increased its use with subsequent products like the Cube, Mac mini and various aluminum iPods. But Jony's team got really serious about machining in 2005 for the iPhone. At that time, they visited various watch manufacturers to see how precise, long-lasting time-keeping products were made.

"We started researching watch companies just to understand machining metals, finishing metals, product assembly," recalled Satzger.[3] What they found was a remarkably high standard of manufacturing. Most importantly, they realized the watch industry used highly machined parts in their high-end products.

While the solution made sense, the Apple investigators also learned that watches are made in relatively small batches. But now the plan gradually evolved to go full bore and use machining as the main manufacturing process for all of Apple's major products.

*   *   *

The "unibody process" is a blanket name for a number of machining operations. Machining in general has long been time- and labor-intensive. It relies on big, slow machines like drills and milling machines, but modern CNC machines have greatly sped up and automated the process.

Traditionally, machining has rarely been used in mass production, which is more likely to rely on fast and efficient methods like stamping and molding to turn out products in the millions. Machining is usually associated with one-shot products or small batches. The prototypes made in Jony's design studio are machined, individually carved using CNC milling machines. In industry, machining has usually been employed only by specialized manufacturers with high standards and deep pockets; think aerospace, defense, high-end watches and designer cars, like the Aston Martin. It is the way to make the best parts possible, the pinnacle of refinement and precision. But it takes time and money.

"Machining enables a level of precision that is just completely unheard of in this industry," said Jony. "We have been so fanatical in the tolerances of how we machine and build these products, in many ways I think it is more beautiful internally than it is externally. I think that testifies to just our care, to how much we care."[4]

Jony saw the unibody process as the key to shrinking the iPhone, iPad and MacBook. In all of these products, a single part forms the back plate and the frame. All the screw bosses are cut into it, for attaching other components, and condensing even more parts into one. Unibody allowed Jony's team to make the iPhone 5 about three millimeters thinner than the iPhone 4S. That may not sound like much, but it trimmed about 30 percent of the thickness off an already thin product.

For a laptop body, the first part of the process is to create a block of extruded aluminum from a billet (a big round tube) of raw aluminum.

The billet is put through a giant hot press that, as if making flat noodles from a ball of dough, creates an extrusion into a sheet of aluminum.

The aluminum sheet then begins a trip through thirteen separate milling operations to get it into its final shape. The metal is cut into rectangular blocks the size of the laptop. It goes into the first CNC machine, where a laser drill creates a series of registration holes that guide the next cutting operation, a rough "hogging out" that removes the majority of the unwanted material.

This step is followed by a series of increasingly precise milling operations that create the finished part. The key caps and input ports are cut out. Screw bosses are cut and internal struts and ribs are shaped.

The next stage is the laser drilling of perforations for indicator lights. The inside of the case, where the light will be located, is milled thin enough to allow a laser drill to perforate minute holes through the metal. The laser drill is extremely accurate and fast, vaporizing metal with each pulse. The perforations are so tiny, the metal appears to be solid from the outside, but in actuality, they're big enough to allow the LED behind to shine through. It's an innovative practice for building magic into the product through precision.

"What is intriguing about that small design detail, is it is a phenomenal piece of design," said designer Chris Lefteri. "An obvious thing to do there would be to make a hole in the metal, insert an LED, and place a piece of plastic over the top. Although that would do the job adequately, what Apple did instead was machine a series of virtually invisible holes in the body of the computer, so that suddenly lights appeared inside the holes. That is a craftsman-like approach to the industrial production process."[5]

The laser drills are also used to make speaker grilles and other small openings, before a blast of fluid clears any debris. Apple uses lasers to etch serial numbers and other technical info onto the case, and may use them to inscribe personal inscriptions on the backs of iPods. Because there's no physical contact with the aluminum, the "drills" used for this process never dull or wear out, and they are easily configured and reconfigured through the CNC controls.

After laser drilling, the unibody is passed to a CNC grinding machine that smooths burrs, rough patches and any surface imperfections. The cases are then "blast finished"—sprayed with dry particulates such as ceramic, silica, glass or metal under high pressure—to give the surfaces a textured, matte finish. Then the part is anodized, clear coated or polished, depending on the finish.

The entire unibody process is very much a trade secret, so Apple reveals few details. How much of the process is automated is not clear, though at least part of the assembly is done by robots. While most Apple products have been assembled by hand by legions of workers, it appears the unibody process may enable the company to shift toward automated assembly.

"There's a lot of focus on robotics and robotic control," said a former mechanical engineer who worked as a liaison among ID, product development and operations, and spent months in the factories. The engineer declined to elaborate, citing confidentiality agreements, but said that many of Apple's products are now primarily made and finished on CNC machines with robots moving parts between machining cycles.[6]

"I have literally seen buildings where as far as the eye can see, where you can see machines carving, mostly aluminum, dedicated exclusively for Apple at Foxconn," said Guatam Baksi, a product design engineer at Apple from 2005 to 2010. "As far as the eye can see."[7]

Unibody Today

The unibody process is revolutionizing high-tech manufacturing. At Apple, the move toward robot workflow has, in a sense, revived Steve

Jobs's long-cherished dream once manifested in his 1980s Macintosh factory in the Bay Area with its automated production line. Machining used to be used strictly for prototyping, and no one employed it on an industrial scale until Jony came along. But others have been quick to recognize its importance. Dennis Boyle, one of the cofounders of IDEO, said machining products on an industrial scale is "a dream for product designers."[8]

"Companies have always traditionally avoided machining because it costs more than other techniques," he said, "but Apple has figured out a way. . . . Apple has proved that if a company invests at the highest level and takes Ive and his team's designs and really sticks to them without compromising on how they look and feel, then it can create products that are so sought after, so beautiful and elegant, that they can make them a success. From a design and engineering point of view, Apple is at the absolute pinnacle of creating products that are as close to flawless as can be done."

Unibody represents a giant financial gamble by Apple. When it started investing seriously around 2007, Apple contracted with a Japanese manufacturer to buy all the milling machines it could produce for the next three years. By one estimate, that was 20,000 CNC milling machines a year, some costing upward of $250,000 and others $1 million or more. The spending didn't stop there, as Apple bought up even more, acquiring every CNC milling machine the company could find. "They bought up the entire supply," said one source. "No one else could get a look in."[9]

This spending on tooling ramped up with the iPhone and iPad, which relied more on machining with each generation. According to Horace Dediu of Asymco, an analyst firm, the original iPhone cost $408 million in equipment investment. But by 2012, as the iPhone 5 and iPad 3 (both unibody products) went to production, Apple's capital expenditures ballooned to even more mind-boggling levels. Apple spent $9.5 billion on capital expenditures, the majority of which was earmarked for product tooling and manufacturing processes. By comparison, the company spent $865 million on retail stores. Thus, Apple spent nearly eleven times as much on its factories as on its stores, most of which are in prime (that is, expensive) real estate locations.[10]

Another manufacturing innovation made necessary by Jony's design desires—in this case, to get razor-thin edges on the 2012 iMacs—is friction welding. Given the thin profile of the iMac, traditional welding methods couldn't be used to join the front and the back. Enter the need for so-called friction stir welding (FSW), a solid-state welding process invented in 1991. It's actually less of a weld than a recrystallization, as the atoms of the two pieces are joined in a super strong bond when a high-speed bobbin is moved along the edges to be bonded, creating friction and softening the material almost to its melting point. The plasticized materials are then pushed together under enormous force, and the spinning bobbin stirs them together. The result is a seamless and very strong bond.

In the past, FSW required machines costing up to three million dollars apiece, so its use was confined to fabricating rocket and airplane parts. More recent advances allowed CNC milling machines to be retrofitted to perform FSW at a much lower cost. That opened the door for Apple, which has many CNC machines at its disposal.

In addition to its other advantages, FSW produces no toxic fumes and finished pieces that require no extra filler metal for further machining, making the process more environmentally friendly than traditional welding.

Greening the Apple?

The new manufacturing methods are driven partly by Jony's desire to make Apple greener. Those desires got a kick start in 2005, when Apple got into a public spat with Greenpeace International. The global environmental campaigner slammed Apple for its lack of a recycling program and its use of a host of toxic chemicals in its manufacturing processes. Steve Jobs dismissed the charges at first but, in 2007, announced a total overhaul of Apple's environmental practices. Since then the company has improved its environmental profile, reducing toxins in manufacturing, including mercury, arsenic, brominated flame retardants and polyvinyl chloride.

In a further attempt to improve its environmental profile, Apple lowered power requirements of many products, earning high Energy Star ratings and gold ratings from the Electronic Product Environment Assessment Tool (EPEAT), which tries to measure products' environmental impact over their lifetime, taking into account energy use, recyclability and how the products are designed and made. Apple has also reduced the size of its packaging, permitting more packages to be loaded into freight and saving fuel. And the newest MacBooks are touted as 100 percent recyclable, while Apple products in general use aluminum and glass, materials that are easily recycled and reused.

Even so, Apple still doesn't get the highest marks from Greenpeace because the company is so secretive. In 2012, Greenpeace gave Apple a score of 4.5 out of a possible 10, putting Apple in the middle of the pack of tech companies (an improvement, actually, as it started at the bottom). Overall, Greenpeace credits Apple with increased environment responsibility but points out that "Apple misses out on points for lack of transparency on GHG [greenhouse gas] emission reporting, clean energy advocacy, further information on its management of toxic chemicals, and details on post-consumer recycled plastic use."[11]

Although Jobs got public credit for the greening of Apple, one source inside the company said a lot of the impetus came from Jony, who was "really stung" by Greenpeace's initial criticism.

"Jony felt that Apple had very positive stories about its environmental impact," the source said. Jony's commitment to not making junk products certainly puts him in a stronger environmental light: His products tend to be used and cherished for years, just the opposite of throwaway products with their more immediate and detrimental environmental impact.

Another criticism cast at Jony and the company concerns the decision to seal many of their products. That means that, in most cases, Apple products aren't user serviceable, as they require special tools and skills even to change a battery. Environmental activists like Kyle Wiens of iFixit point out a sealed device is more likely to be junked than one that can be more easily repaired by the nonprofessional.

The iPad, Wiens said, is "deeply immoral. It's glued shut and the battery will be inoperable after five hundred charge-discharge cycles. It is expressly designed to be thrown away. Reparability isn't a concern and so designers aren't going to design that in."[12]

Apple insiders disagree. They point out Apple's products are certainly designed to be repaired, though not by their owners. "Apple has a special service process," Satzger explained. "Not a lot of other companies have the ability to service their own products, so they design them to be serviced by places like Best Buy."

Satzger argued that that means Apple equipment is in fact more serviceable. "For repair work, Apple takes back all its products and handles servicing through the company's stores. Repairs and servicing are taken into account from the outset of a product's design. . . . Apple's service process is exquisitely refined for their own products."[13]

Apple, as one of the world's richest and most powerful companies, has clearly taken a leadership role in manufacturing. If their commitment to their global workforce and to environmental concerns remains less certain, it's clear that Jony Ive will have a voice in shaping those policies into the foreseeable future.

CHAPTER 13

Apple's MVP

[Jony Ive] has more operational power than anyone else at Apple except me. There's no one
who can tell him what to do, or to butt out. That's
the way I set it up. —STEVE JOBS

Steve Jobs had surgery for a pancreatic tumor in July 2004. As he was recovering from his first bout with cancer, he asked to see two people. One was his wife, Laurene Powell Jobs; the other was Jony Ive.

After nearly eight years of working together almost daily, Jony and Jobs had a special and intimate relationship. The pair had been nearly inseparable, attending many of the same meetings, eating lunch together and spending afternoons at the studio going over future projects.

Jobs's first surgery didn't fully cure him and he later underwent a second round of surgery, taking a leave of absence from Apple to undergo a liver transplant in Memphis, Tennessee, in May 2009. Jobs flew home on his private jet with his wife, where he was met by Jony and Tim Cook at San Jose Airport. The question of Apple's future was very much in the air, as the announcement of Jobs's leave had led many in the press to predict that Apple was doomed without him. It seemed to be the consensus of the punditocracy that the fate of Apple rested solely on Jobs's shoulders.

Jony drove Jobs home from the airport and confided on the journey that he was disturbed by newspaper opinion pieces that staked Apple's survival to Jobs.

"I'm really hurt," Jony told Jobs. He was worried about Jobs's health, and the health of the company they both loved. As Jony told Jobs's biographer Walter Isaacson, the perception that Jobs was the engine of Apple's innovation was damaging, he said. "That makes us vulnerable as a company," Jony said.[1]

That Jony's ego wasn't always sublimated to Jobs's and to Apple is hardly surprising. On another occasion, Jony also complained about Jobs's habit of stealing his ideas. "He will go through a process of looking at my ideas and say, 'That's no good. That's not very good. I like that one,'" Jony told Isaacson. "And later I will be sitting in the audience and he will be talking about it as if it was his idea. I pay maniacal attention to where an idea comes from, and I even keep notebooks filled with my ideas. So it hurts when he takes credit for one of my designs."[2]

Nonetheless, Jony acknowledged, he could have never accomplished what he has without Jobs. "In so many other companies, ideas and great design get lost in the process," he said. "The ideas that come from me and my team would have been completely irrelevant, nowhere, if Steve hadn't been here to push us, work with us and drive through all the resistance to turn our ideas into products."[3]

During 2011, with Jobs on leave that, in the end, proved to be his last, a rash of news stories claimed that Jony was threatening to leave Apple at the end of a three-year stock deal. Jony and his wife reportedly wanted their twin boys to be educated in Britain. The UK newspaper the *Guardian* published a story about Jony's impending departure, giving it the headline "Apple's Worst Nightmare"[4]; London's *Sunday Times* ran a feature saying that Jony was "at loggerheads" with Apple over his desire to move from Cupertino back to Britain, where he owned a home in Somerset. One story suggested Jony would commute between Britain and California.[5]

Threatening was probably the right word; Jony didn't quit. According to an unnamed friend quoted in the British newspaper article, Jony was "just too valuable to Apple and they told him in no uncertain terms that if he headed back to England he would not be able to sustain his position with them." To cement his connection with Apple, the company reportedly paid Jony a $30 million bonus and offered him shares worth a further $25 million. At the time, Jony's personal fortune was estimated at $130 million.

In retrospect, the facts suggest Jony had no intention of moving. He sold a mansion he owned in Somerset close to his parents because he wasn't using it. Jony appeared more committed to Apple than ever.

Jony regularly gets calls from other companies and headhunters, offering him lucrative opportunities to design everything from cars to shoes. But he's emphatically said no to the question of whether he would leave Apple. "The thing is, you could transplant me and this design group to another place and we wouldn't work at all," he said.[6]

On August 24, 2011, Apple announced that Steve Jobs was resigning as CEO, but would remain with the company as chairman of the board. Tim Cook officially took over the day-to-day running of the company.

The news shouldn't have been a surprise but it was. Jobs had been on medical leave since January, and he was obviously a very sick man, appearing emaciated during his few public appearances that year. Even in the face of such a harsh reality, however, everyone found it difficult to imagine an Apple without Steve Jobs.

Many pundits weighed in, arguing that Jony should take over. He had a public profile (Cook did not) because of all the promotional videos he'd appeared in and the awards he'd scooped up. But few serious Apple observers pegged him as the next CEO—not even Jony himself. As one former member of his design team said of Jony's attitude toward being CEO, "Jony doesn't care about all those aspects of running a company." Explaining that Jony had no interest in the business side of Apple—just as he had hated the business side at Tangerine—the designer concluded, "He just wants to focus on ID."

"All I've ever wanted to do is design and make; it's what I love doing," Jony told one interviewer. "It's great if you can find what you love to do. Finding it is one thing but then to be able to practise that and be preoccupied with that is another."[7]

As the master of Apple's global supply chain, Tim Cook was, in fact, much the logical successor. In just thirteen years, Cook had constructed a complex apparatus that allowed the company to build superb gadgets—albeit of Jony's design—at unparalleled speed, volume, efficiency and profitability. Cook might not possess Jobs's charisma, but he was a logistics titan who had been effectively running the company ever since Jobs had gone on his most recent medical leave. He also had experience, serving as interim CEO during Jobs's absences in 2004 and 2009. According to Apple insiders interviewed for this book, Cook is generally affable, a consensus player who wants everyone to buy in, which makes him easier to work with than willful personalities like Jobs.

Jobs died just over a month after his resignation, on October 5, 2011. The man who once said in a speech at Stanford University that "death is very likely the single best invention of life" was about to be eulogized at just fifty-six. His funeral two days later in Alta Mesa was attended by just four of his colleagues from Apple: vice presidents Eddy Cue (software) and Katie Cotton (communications), CEO Cook and Jony. The memorial service held for Jobs at Stanford University ten days after that, though still private, drew former U.S. president Bill Clinton, former vice president Al Gore, Bill Gates, Google CEO Larry Page, U2 frontman Bono and News Corp CEO Rupert Murdoch. Jony had a chance to publicly mourn his mentor and friend three weeks after Jobs's death. At a staff memorial service held on Apple's Cupertino campus, Jony gave the most heartfelt (and humorous) speech of the day, filling his eight-minute tribute to Steve—his "best and most loyal friend"—with personal anecdotes. By turns Jony was funny, touching and insightful, as he described Steve's passion and enthusiasm, his sense of humor and his great joy in doing things right.

Jony's eulogy started with an aside. "Steve used to say to me a lot, 'Hey Jony, here's a really dopey idea.' And sometimes they were really dopey. Sometimes they were truly dreadful. But sometimes they took the air from the room, and they left us both completely silent."

Jony remembered that Jobs "constantly questioned, 'Is this good enough? Is this right?'" He saw Jobs's great triumph as "the celebration of making something great for everybody, enjoying the defeat of cynicism . . . the rejection of being told a hundred times, 'You can't do that.'"

Jony closed by telling the massed Apple employees at the memorial, "We worked together for nearly fifteen years—and he still laughed at the way I said 'aluminium.' For the last two weeks, we've all been struggling to find ways to say good-bye. This morning, I simply want to end by saying, 'Thank you, Steve. Thank you for your remarkable vision which has united and inspired this extraordinary group of people—for all that we have learned from you and for all that we will continue to learn from each other, thank you Steve.'"

Apple's Fortunes Magnify

The day before Steve Jobs died, Cook debuted the iPhone 4s at the Yerba Buena Center for the Arts in San Francisco. There was an empty seat for Jobs marked "Reserved"; Jony was notably absent.

The iPhone 4s was Jony's third-generation design, based on Richard

Howarth's early Sandwich concepts, and the device was the most advanced iPhone yet. Though it shared the same physical appearance as its predecessor, the iPhone 4, it had much-improved guts, and was a marvel of engineering. When the new iPhone went on sale on October 14, some critics called it overhyped and more of the same, but to judge from sales, the public disagreed. The first weekend saw a record-breaking debut, with four million units sold, and the iPhone 4s quickly become the world's best-selling smartphone.

The importance of Apple's first successful post-Jobs product launch did not go unnoticed on Wall Street. Apple's stock began to soar. A share of Apple's stock sold for $407.61 on January 3, 2012, reflecting a balance sheet that contained over $100 billion in cash, a sum that grew by the day. By the end of January, a single share of Apple cost $447.61.

Apple was riding high, having surpassed ExxonMobil as the most valuable publicly held company in the world.

Sir Jony Ive

The year 2012 began auspiciously for Jony Ive, as it had for Apple, despite Jobs's passing. Jony was named a Knight Commander of the British Empire (KBE) in the Queen's New Year Honours List, for services to design and enterprise. It was the second time he had been recognized in the honors list, having been made a Commander of the British Empire (CBE) in 2005. The second highest order of chivalry, the KBE entitled its new bearer to style himself Sir Jonathan Ive.

Jony described the honor as "absolutely thrilling" and said he was "both humbled and sincerely grateful." In a rare interview with the *Daily Telegraph*, he said he was "the product of a very British design education," adding that, "even in high school, I was keenly aware of this remarkable tradition that the UK had of designing and making. It's important to remember that Britain was the first country to industrialize, so I think there's a strong argument to say this is where my profession was founded."[8]

Phil Gray, Jony's first boss at Robert Weavers Group, met up with Jony during the 2012 Summer Olympics in London. "When I asked Sir Jony what was it like being a knight of the realm, he replied: 'You know what? Out in San Francisco it means absolutely nothing. But back in Britain it is a burden.'"[9] Jony is referring to Britain's strong class divisions. He's no longer an everyman. He's been elevated to a Knight of the Realm, and it embarrasses him.

Jony used his visit to London to talk to some design students from Northumbria, his alma mater. He arranged for the temporary closure of an Apple store in London, inviting the students in for a private talk. "Jony likes to get his opinions across; there is no question about that," said a source. "But it is also important to him to give students some support. I guess that is his way of giving something back."

As Northumbria's most famous grad, Jony is regarded by his alma mater as a very valuable asset. Awarded the status of a visiting professor, he occasionally returns to give lectures. He features heavily in Northumbria's marketing materials, but the college, protective of their special relationship, declined to talk about Jony or to release any information about his studies there.

He makes frequent visits to his homeland, staying in London three or four times a year. He has been seen at the London Fashion Week and attends the annual Goodwood Festival of Speed, a race meet for fans of fast, exotic cars. He's served as a judge at Goodwood, where he's been photographed with fellow designer Marc Newsom and composer Nick Wood, two of his best friends. The three often attend one another's functions. Jony is also friends with Paul Smith, the British clothes designer, whom he presented with a giant pink iPod nano for his birthday.

Though design is sometimes thought of as a lonely, isolating process, Jony travels the world regularly. Although he will sometimes spend weeks with suppliers in Asia, on most trips, he's in and out quickly. In 2013, he traveled to Amsterdam for a day, during which he went aboard Steve Jobs's boat (designed by Philippe Starck and custom-built in Holland) and opened the new Apple store there.

In 2012, Jony and his wife and twin sons upgraded to a new San Francisco home, purchasing a seventeen-million-dollar spread on San Francisco's "Gold Coast," also known as Billionaire Row. Despite his image as a soft-spoken everyman in jeans and T-shirt, he's often photographed at exclusive venues with other well-suited high rollers. When at home in San Francisco, he's been known to attend the symphony and he socializes with the Silicon Valley elite. He's been photographed at celebrity dinners with valley bigwigs such as Yahoo CEO Marissa Mayer, Twitter CEO Dick Costolo and the CEOs of Yelp, Dropbox and Path.

Occasionally, Jony gets involved in side projects. He designed some striking Soundstick speakers for Harman Kardon, which are part of the Museum of Modern Art's permanent collection. In 2012, he designed on commission a one-shot camera for Leica, which was to be auctioned for charity. Jony and Jobs were both fans of the storied camera maker, and when announcing the iPhone 4, Jobs compared it to "a beautiful old Leica camera."[10]

By all appearances, Jony remains committed to Apple, despite occasional rumors to the contrary. He's also reportedly working on a monograph of his work at Apple.

Apple Carries On

Even without Steve Jobs at hand to challenge him, Jony in 2012 remained a busy and engaged man. In March, Apple announced the third-generation iPad, called simply the "new iPad." It would have the strongest launch of all previous iterations of the device, selling three million units over the first weekend.

The iPhone 5 was announced several months later, in September 2012. Redesigned with a bigger, four-inch retina display, the iPhone 5 was "the most beautiful consumer device that we've ever created," said Phil Schiller. Preorders of the new phone topped two million in the first twenty-four hours after the announcement; upon its release on September 21, the phone set a new record with weekend sales of over five million units, and demand exceeded supply for several weeks.

One month after the iPhone 5, Apple announced the release of the second new iPad within the same year, along with its little cousin, the iPad mini, a tablet computer with a smaller, 7.9 inch display. In November, the two new tablets were released simultaneously in thirty-four countries and, between them, sold three million units in just three days. The surprise iPad was unusual because Apple's normal release schedule didn't shift significantly. Products came no quicker than they had before.

The new products proved a further boon to Apple's stock. Within days after their release, Apple's share price rose over 12 percent, from $505 to $568, and continued to climb thereafter.

Apple seemed to be gaining strength—there was no sign the company was suffering from the loss of Steve Jobs. On the contrary, it was a period of incredible financial and creative fecundity. But on October 29, 2012, Apple announced a surprise executive reshuffle.

In a press release that seemed calculated to obfuscate what was really going on behind the scenes, Apple announced a major shift of executive offices. To put it more bluntly than the Apple communications did, Scott Forstall was fired from his role as head of iOS and Jony was promoted to overall creative head.

Jony would maintain his position of senior vice president of ID, but henceforth would also "provide leadership and direction for Human Interface across the company."[11] In other words, Jony would be in charge of the all-important product interfaces in both hardware and software, a role previously fulfilled by Steve Jobs. "Jony has an incredible design aesthetic and has been the driving force behind the look and feel of our products for more than a decade," said Cook in a follow-up e-mail to employees. "The face of many of our products is our software and the extension of Jony's skills into this area will widen the gap between Apple and our competition."[12]

When parsed carefully, Cook was saying, in short: The man who had long been responsible for setting the direction for Apple's hardware would now be given the power to reimagine its software as well. Although Apple chose to neither confirm nor deny them, rumors had it that Forstall was ousted over fallout with Jony over user-interface design. The departure of Forstall and Jony's increased responsibilities strongly suggested that Forstall lost a power struggle.

A key contention concerned Forstall's fondness for skeuomorphic design; that is, graphic interfaces that resemble real-world objects. Apple's user-interface conventions under Forstall tended to look like their real-life counterparts. Virtual wooden shelves were used to display eBooks in the iBookstore app; Apple's Podcast app looked like a reel-to-reel tape recorder; iOS's multiplayer gaming service, Game Center, was styled like a Vegas casino table. Faux leather and wood-grain patterns had found their way into many of Apple's most popular apps.

Such skeuomorphic design allows neophyte users to be immediately familiar with an unfamiliar device, operating on the assumption that nothing is simpler than an interface that works exactly like objects do in the real world. The original Macintosh desktop computer, for example, was conceived as a skeuomorphic version of an office desktop as seen from above. Because everybody knew how the items on a traditional desk were used in the physical world, that knowledge could be implicitly transferred to its digital counterpart.

More recently, however, Apple had heard loud criticisms concerning its use of "tacky" skeuomorphic elements. According to some, visual references to obsolescent office furniture and audio equipment were beginning to look dated and out of place. Forstall, after Jobs's death, was reportedly Apple's major champion of skeuomorphic design, which put him in the line of fire not only in the eyes of external critics but from some within Apple too.

Jony Ive was never a fan of skeuomorphism, according to one unnamed Apple designer speaking to the *New York Times*.[13] In an interview with the UK's *Telegraph*, Jony visibly "winced" when the subject came up, but refused to be drawn into a detailed discussion. "My focus is very much working with the other teams on the product ideas and then developing the hardware and so that's our focus and that's our responsibility," he said. "In terms of those elements you're talking about, I'm not really connected to that."[14] Cook's reshuffle corrected that.

Apple's management shake-up represented a major design shift in software and, by the time iOS 7 was released in 2013, most of Forstall's skeuomorphic references were nowhere to be seen.

The mobile software was flat and modern looking. Gone were references to felt and leather, as well as 3-D effects like highlights and shadows. "No virtual cows were harmed in the making of this," joked Craig Federighi, senior vice president of software engineering, as he showed off iOS 7's calendar app during the launch event. He added that other apps were cleaner, too, because "We just completely ran out of green felt. And wood too. This has to be good for the environment."[15] The iOS 7 design was minimal, bearing a curious resemblance to the phony operating system that Jony's group used when designing the iPhone and iPad hardware in the mid-2000s. They had the same flat look, and some of the icons were very similar. The reversion back to the mock-up OS hinted at the animosity between Jony and Forstall, suggesting that Jony's instincts for software design had been downplayed to Forstall's for several years.

On the other side of the calculation, Jony's overhaul of iOS was consistent with his approach to hardware. Jony's hardware has always been about bare, utilitarian minimalism. He disdains decoration—as he says, every tiny screw is there for a reason—and his goal is to make design disappear. In contrast, skeuomorphism is about making software look like something it isn't, like a roulette table or a yellow legal pad, and decoration is essential. Skeuomorphic software is the opposite of Jony's minimalist hardware. One strips away everything that isn't necessary; the other puts it back in.

This paradox within Apple ended with iOS 7. With the ornamentation taken out, Jony's software was in sync with his hardware, stripped to their essentials. In addition, iOS 7 showed good design taste, especially in the use of typography. It featured a typeface called Helvetica Neue, a fine, detailed Swiss-designed typeface that's enabled by the detailed retina displays of Apple's latest devices. The entire operating system was infused with a deep appreciation for print-graphic design.

As Apple's mobile devices mature, the shift to put Jony in charge of hardware and software is hugely significant. Jony and his design team will continue to improve the hardware, but changes are likely to be incremental, not fundamental. There are only so many ways a thin glass rectangle can evolve. These days, the design frontier is software, not hardware.

Sally Grisedale, the fellow Brit who worked with Jony in the mid-nineties, said Jony has always been adept at software. "[Jony and OS design] are a perfect fit," she said. "It was always about the hardware software integration. . . . This whole piece of hardware-software interaction is the most exciting arena and he has sort of been leading the way for years. This is not new for him; he has always thought that way and it was just a question of scale and scope. Perfect fit."[16]

Larry Barbera, one of Jony's old design colleagues, also thought Jony was well equipped to refresh Apple's software. But he also pointed out the need for him to build relationships with programmers. Despite being immersed in the software side of the business for years, Barbera said, "Jony needs to evangelize the software folks by creating a vision for all to buy in to. I'm sure that half of Jony's battles will be in winning over the hearts and minds of the software folks."[17]

Jony's appointment to software came at a crucial time. Apple's competitors are catching up, as the Android continues to mature and attract the kind of user who likes more control and more choices. Microsoft's Windows 8 won plaudits for its clean, ambitious touch interface. "This is a defining moment, where hardware fulfills its promise and simply gets out of the way," wrote Alex Schleifer, design and creative director at Say Media, a San Francisco advertising company. "A shape of glass existing solely to contain an experience. The user interface will be how we remember a device, fondly or not. The way it looks and reacts. It will live in our cars and living rooms, become part of the architecture, cover our landscapes. It will affect the media we consume, the way we look at the world, and how we learn and communicate. Here's to the age of the user interface."[18]

In an interview published in *Bloomberg BusinessWeek*, Cook shed little light on the top-level changes and direction at Apple, beyond deflecting inquiries with some standard Cook-style remarks: "Creativity and innovation are something you can't flowchart out. You know, small teams do amazing things together. Collaboration is essential for innovation."

He did, however, go as far as to opine warmly that "I don't think there's anybody in the world that has better taste than Jony Ive does." Cook added, "Jony and I both love Apple. We both want Apple to do great things."[19]

"Jony Is Irreplaceable"

Before he died, Jobs revealed the degree to which he empowered Jony inside the company. "He has more operational power than anyone else at Apple except me," Jobs said. "There's no one who can tell him what to do, or to butt out. That's the way I set it up."[20]

Jobs didn't explain exactly what he meant. According to Apple's organization chart, Jony reports to Cook; yet, according to Jobs, Cook can't tell him what to do. If such an arrangement seems unusual, it's because Jony has enormous operational clout. His IDg, as the most powerful in the company, calls the shots with engineering and manufacturing. The company has invested billions in Jony's demanding manufacturing methods. Budget and feasibility has never been a factor as Cook's operations group implemented Jony's IDg's designs.

For many years, Jony had transcended his role as IDg head, particularly by growing into his unusual partnership with Steve Jobs. Jobs saw Jony as a true collaborator and innovator.

"He understands business concepts, marketing concepts. He picks stuff up just like that, click," Jobs told Isaacson. "He understands what we do at our core better than anyone. If I had a spiritual partner at Apple, it's Jony. Jony and I think up most of the products together and then pull others in and say, 'Hey, what do you think about this?' He gets the big picture as well as the most infinitesimal details about each product. And he understands that Apple is a product company. He's not just a designer."[21]

Friends and ex-friends alike acknowledge that Jony has carefully guarded his image as a soft-spoken English gentleman—but that's he is also a practiced corporate player. Generous and protective as he may be of his IDg team, Jony has a healthy ego and isn't shy about claiming personal credit for ideas or innovations. His jousts with Forstall and Rubinstein reveal another more aggressive facet of his character: He isn't afraid to take on executive colleagues. To judge from the outcomes of the corporate contretemps that have come to light, Jony Ive possesses both the determination and the corporate firepower to prevail when he chooses to engage in such turf battles.

In the creative sphere, there's little doubt that Jobs groomed Jony as his absolute successor, though without the CEO title. Cook would keep the trains running on time, but Jony, as the product champion, was endowed with operational muscle across the company.

The result has been that, in many ways, things have stayed the same. "We're developing products in exactly the same way that we were two years ago, five years ago, ten years ago," said Jony. "It's not that there are a few of us working in the same way: there is a large group of us working in the same way."[22]

Jony clearly seeks to maintain Jobs's values: To Jony, as it was to Jobs, making "great products" is much more important than the balance sheet. "Our goal isn't to make money," Jony told a surprised audience at the British Embassy's Creative Summit in July 2012. "Our goal absolutely at Apple is not to make money. This may sound a little flippant, but it's the truth. Our goal and what gets us excited is to try to make great products. We trust that if we are successful people will like them, and if we are operationally competent we will make revenue, but we are very clear about our goal."[23]

Jony explained that he learned this lesson from Jobs when Apple was poised to go under. "Apple was very close to bankruptcy and to irrelevance [but] you learn a lot about life through death, and I learnt a lot about vital corporations by experiencing a non-vital corporation," he told the conference. "You would have thought that, when what stands between you and bankruptcy is some money, your focus would be on making some money, but that was not [Steve Jobs's] preoccupation. His observation was that the products weren't good enough and his resolve was, 'We need to make better products.' That stood in stark contrast to the previous attempts to turn the company around."

Jony is also committed to maintaining Jobs's renowned focus. Jobs always said that focus isn't a question of saying yes to projects; it's saying no. Under Jony's guidance, Apple has remained highly disciplined in "saying no" to products that are "competent" as opposed to "great."

"We have been, on a number of occasions, preparing for mass production and in a room and realized we are talking a little too loud about the virtues of something. That to me is always the danger, if I'm trying to talk a little too loud about something and realizing I'm trying to convince myself that something's good," he said.[24]

Clive Grinyer, Jony's first business partner, expressed confidence that Jony would catapult Apple to even greater heights.

"Jony never was just the designer," said Grinyer. "He always played a much more strategic role at Apple. That includes also the user interface, for which he also helped make the decisions. . . . Jony is now in the strategic position. I always felt very optimistic about Apple, because so much of Apple's success has been due to Jony. Steve unlocked Jony. Steve took Jony away from printer lids and gave him the job he was capable of. . . . Steve gave Jony the confidence to bring out his innate design talent and create amazing products. And from now he will carry on as normal.

"Apple was already a pretty amazing company, but the level they have reached in the last ten years is because they have had Jony, empowered by Steve, to produce that incredible panel of work."[25]

Grinyer took his argument one step further. "Believe it or not, Jony's leaving would be worse for Apple than Jobs's leaving. Jony is irreplaceable. If he were to go, to get another design leader with that sense of humanity, vision, calmness and ability to keep the team together, would be impossible. Apple would become something different."[26]

As Jony has said, "A big definition of who you are as a designer, it's the way you look at the world. And I guess one of the curses of what you do, is you are constantly looking at something and thinking, 'Why? Why is it like that? Why is it like that and not like this?'"[27]

Very likely, as a man who see himself "constantly designing," Jony Ive will continue to do precisely that at Apple into a future that will feature new designs, new products—and some happy surprises.

Reading the Weft and Warp

There are intriguing symmetries and continuities to be observed in Jony's life and career. His father was an education reformer whose work directly affected his son's design schooling. Jony's first school projects were futuristic phones in white plastic. A prototype tablet got him his job at Apple.

Jony's career has been marked by connections (Grinyer, Brunner) and coincidences (Brunner moving to Apple). But if there have been a few lucky breaks, it is equally true that he made his own luck.

Jony was always crazy for design. As a kid, he wasn't just talented, he was a design prodigy. He was aided by his dad, whose personal passion was instilling a passion for design in Britain's school kids. His education at Newcastle Polytechnic was very hands-on, likewise developing his ongoing interest in making, which would be apparent again later in his love of making prototypes and pioneering new forms of mass production.

Jony's early experience as a consultant at Tangerine built in him a consultant's mentality and workflow, which Jony brought to Apple, where the design studio operates like a consultancy, only within a large corporation.

"I had been concerned that moving away from working independently for a number of clients on a broad range of products would be difficult," he once said. "Surprisingly this has not been an issue, as we are really designing systems that include so many different components— headphones, remote controls, a mouse, speakers as well as computers."[28]

Some of Jony's first work at Apple, including the Newton MessagePad and the Twentieth Anniversary Mac, were harbingers. Jony hired many of the core members of his team when Apple was in difficulty, and he protected and nurtured them during Apple's darkest days. It was this team that would contribute so much to Apple's smash hits in the years that followed.

His collaboration with Steve Jobs, starting with the iMac, became nothing less than one of the most fruitful creative partnerships in history. Together they reset Apple's engineering-driven culture and created a much more tightly integrated design-driven approach, where "design" (meaning creative engineering, whether it's hardware, software or advertising) permeated everything the company does.

The products that followed sent Jony deep into new materials and manufacturing methods, driven by his desire to always find a better way. The iPod was a product of Jony's simplification philosophy. It could have been just another complex MP3 player, but instead he turned it into the iconic gadget that set the design cues for later mobile devices. Two more delightful innovations, the iPhone and iPad, were products of thinking differently, of creative engineering at work in rational problem solving on many levels.

The unibody process is Jony's genius for simplification applied to process manufacturing. With his work in machining, it's not unreasonable to say he achieved some pinnacle of making on an industrial scale. No wonder D&AD, in 2012, named Apple the best design studio and best brand of the past fifty years. D&AD awards are the Oscars of the industry, and Jony has ten of them, more than any other designer.

The process of simplification is design 101, a mind-set that every design student is taught in school. But not every student adopts it, and it's rarely applied with the ruthless discipline practiced by Jony. Indeed, if there's such a thing as a single secret to what Jony Ive does, it is to follow slavishly the simplification philosophy. That approach has accounted for many of the major breakthroughs, as well as for some products that failed and others that Apple hasn't released. Caring enough to commit the enormous time and effort to get something right has also been Jony's hallmark, from his earliest college projects onward.

Jony's ultimate goal is for his designs to disappear. The shy boy from Chingford is happiest when the user doesn't notice his work at all. "It's a very strange thing for a designer to say, but one of the things that really irritates me in products is when I'm aware of designers wagging their tails in my face," he said. "Our goal is simple objects, objects that you can't imagine any other way. . . . Get it right, and you become closer and more focused on the object. For instance, the iPhoto app we created for the new iPad, it completely consumes you and you forget you are using an iPad."[29]

Andrew Hargadon, a design and innovation professor at the University of California at Davis, who worked in Apple's design studio before Jony took over, said not only has Jony made computers and smartphones indispensable, he's also driving a cultural shift for better design.

"When the iMac first came out in jelly bean colors, so many other different products came out following that lead. There were staplers in six jelly bean colors. The iMac turned consumers into design aficionados to a much greater degree than they were before," Hargadon said. "That's probably the single greatest effect, that we nowadays expect many things to have better designs. Because of Apple, we got to compare crappy portable computers versus really nice ones, crappy phones versus really nice ones. We saw a before-and-after effect. Not over a generation, but within a few years. Suddenly 600 million people had a phone that put to shame the phone they used to have. That is a design education at work within our culture."[30]

Jony's challenge now is to keep Apple fresh and innovative. In the dark days before Jobs returned, the biggest risk was not being risky. If Apple hadn't taken risks—some of which paid off big—it might have been out of the game. Today, as an immensely successful corporation with established territories and a dominant place in several markets, the danger has passed and, with each generation of Apple's products, the generational leaps grow more incremental and pose fewer risks.

But Apple's success—and the continuity that Jony has brought to the company—has come to mean that the customer can now almost anticipate what a product is going to look like; the shock of the new is gone. "Apple has created a very careful brand DNA, which has however become a noose around their necks, from which they cannot shift," warned Professor Alex Milton. "Apple has gone from being the alternative to the mainstream."[31]

Milton sees this as a tension for Jony, given that more recent graduates of the sort of design schools he attended are now rejecting that aesthetic. "Ive is the establishment," asserted Milton. "The challenge for Ive is, can he reinvent himself, or is he stuck in time?

"Apple has to find a new language, and the challenge is, what is, that going to be? I have confidence that Jony Ive has the wherewithal to drive the next step for Apple, but this is by far the most difficult point."

ACKNOWLEDGMENTS

First and foremost, I'd like to thank my literary agent, Ted Weinstein, for pushing me to do this book. The team at Portfolio, including the boss, Adrian Zackheim, have been great. Natalie Horbachevsky and Brooke Carey did an excellent job editing, and Hugh Howard performed a postproduction miracle editing the manuscript and shaping the narrative.

Jose Garcia Fermoso and my brother, Alex Kahney, provided invaluable help finding, contacting and interviewing sources.

I'd like to thank John Brownlee for helping me with the writing and running of the Cult of Mac blog in my many long absences. Thanks also to my colleagues on the blog: Charlie, Buster, Killian, Alex, Rob and Erfon, for minding the shop and doing a great job.

The book has benefitted greatly from the reporting of others, in particular Paul Kunkel's *AppleDesign*, Luke Dormehl's *The Apple Revolution* and Walter Isaacson's biography of Steve Jobs.

SECRECY AND SOURCES

There's a T-shirt you can buy at the Company Store on Apple's campus. It says, "I visited the Apple campus. But that's all I'm allowed to say." That just about sums it up when reporting on Apple.

Trying to persuade people to talk about the company isn't easy. Apple people don't talk, even about things that happened thirty years ago. The company is so secretive, that divulging anything—anything at all—is a firing offense. Everyone associated with the company—employees, contractors, partners—has signed a stack of nondisclosure agreements, which threaten not just termination but prosecution to the fullest extent of the law. Employees are mum about current product plans, which is understandable, but they won't talk about old projects either. The secrecy extends to every single thing Apple does, but especially applies to its internal processes, which it considers industrial trade secrets. Apparently, knowing how Apple conducts meetings, for example, could give competitors a leg up.

Apple is the ultimate need-to-know culture. It operates like a spy organization. Staffers are told the absolute bare minimum to do their (highly specialized) jobs. Only a handful of executives and senior VPs have the whole picture, and often they don't know what happened in other departments outside their own, or what happened in the rank and file.

Secrecy is so steeply embedded in Apple's culture, keeping mum is as natural as breathing. Apple employees live in an Apple bubble. They do not attend conferences or give talks, and they barely circulate in Silicon Valley's professional or social circles. Friends of employees know not to ask their Apple friends about work. If the subject comes up, it's met with an apologetic smile. They won't even share with their spouses. One female designer interviewed for this book said she and her husband are especially careful not to talk about work because everyone expects them to; they work extra hard not to.

We contacted more than two hundred people for this book, mostly current Apple staffers or those who had recently left the company. Some were willing to talk on the record, but many wanted to keep their names out of print. Apple did not respond to several requests for comment.

Nonetheless, my research partners and I did get a number of people to talk, on and off the record, about Apple, Jony Ive and their unique work culture. Notably, we got some major players, including some who have worked closely with Jony for decades. They took us inside the studio and inside the minds behind Apple, in an unprecedented manner. The information they provided, and the details they were able to reveal about the company over many years, were immeasurably useful.

Thanks to their interviews, coupled with extensive research, and the available videos, transcripts, launch archives, books, articles and the Apple product output itself, we offer the fullest picture available of the true events behind Jony Ive's career and influence at Apple.

NOTES

CHAPTER 1
School Days

1. London Design Museum, interview with Jonathan Ive, http://designmu seum.org/design/jonathan-ive, last modified 2007.
2. Interview with Ralph Tabberer, January 2013.
3. Ibid.
4. Design and technology curriculum of UK schools, http://www.education.gov.uk/schools/toolsandinitiatives/a0077337/design-and-technology-dt, updated November 25, 2011.
5. Interview with Malcolm Moss, January 2013.
6. Interview with Ralph Tabberer, January 2013.
7. Rob Waugh, "How Did a British Polytechnic Graduate Become the Design Genius Behind £200 Billion Apple?" *Daily Mail*, http://www.dailymail.co.uk/home/moslive/article-1367481/Apples-Jonathan-Ive-How-did-British-polytechnic-graduate-design-genius.html, last modified 3/19/13.
8. John Coll and David Allen (Eds.), *BBC Microcomputer System User Guide*. http://regregex.bbcmicro.net/BPlusUserGuide-1.07.pdf
9. Waugh, "How did a British polytechnic graduate become the design genius behind £200 billion Apple?"
10. Shane Richmond, "Jonathan Ive Interview: Apple's Design Genius Is British to the Core," *Telegraph*, http://www.telegraph.co.uk/technology/apple/9283486/Jonathan-Ive-interview-Apples-design-genius-is-British-to-the-core.html, May 23, 2013.
11. Walter Isaacson, *Steve Jobs* (Simon & Schuster, 2011), Kindle edition.
12. Paul Kunkel, *AppleDesign*, (New York: Graphis Inc., 1997), p. 253.
13. David Barlex, "Questioning the Design and Technology Paradigm," Design & Technology Association International Research Conference,

April 12–14, 2002, pp. 1–10, https://dspace.lboro.ac.uk/dspace-jspui/bitstream/2134/3167/1/Questioning%20the%20design%20and%20technology%20paradigm%20.pdf.

14. Mike Ive OBE, keynote address 1, "Yesterday, Today and Tomorrow," NAAIDT Conference 2003 Wales, Developing Design and Technology Through Partnerships, archive.naaidt.org.uk/news/docs/conf2003/MikeIve/naaidt-03.ppt.

15. E-mail from a former schoolmate, October 2012.

16. Interview with Craig Mounsey, March 2013.

17. Interview with Dave Whiting, September 2012.

18. Interview with Phil Gray, January 2013.

19. Ibid.

20. "Provisional GCE or Applied GCE A and AS and Equivalent Examination Results in England," http://www.education.gov.uk/researchandstatistics/datasets/a00198407/a-as-and-equivalent-exam-reults-2010-11.

21. John Arlidge, "Father of Invention," *The Observer*, http://observer.guardian.co.uk/comment/story/0,6903,1111276,00.html, December 21, 2003.

CHAPTER 2

A British Design Education

1. Northumbria University, About Us page, http://www.northumbria.ac.uk/sd/academic/scd/aboutus/.

2. Interview with David Tonge, January 2013.

3. Interview with Paul Rodgers, October 2012.

4. Interview with Craig Mounsey, March 2013.

5. Design for Industry, BA (Hons), Course Information, 2013 entry, http://www.northumbria.ac.uk/?view=CourseDetail&code=UUSDEI1.

6. Industrial Placement Information Handbook, Northumbria University School of Design, Placement Office, 2011–2012, http://www.northumbria.ac.uk/static/5007/despdf/school/placementhandbook.pdf.

7. Octavia Nicholson, "Young British Artists," from Grove Art Online, Oxford University Press, http://www.moma.org/collection/theme.php?theme_id=10220.

8. Interview with Penny Sparke, September 2012.

9. Interview with Alex Milton, October 2012.

10. Ibid.

11. Carl Swanson, "Mac Daddy," *Details*, February 2002, volume 20, issue 4.

12. Nick Carson, first published in Issue 5 of *TEN4*: Jonathan Ive: http://ncarson.wordpress.com/2006/12/12/jonathan-ive/, Jonathan Ive in conversation with Dylan Jones, editor of British *GQ*, following his award of honorary doctor at the University of the Arts London, November 16, 2006.

13. Ibid.

14. Rob Waugh, "How Did a British Polytechnic Graduate Become the Design Genius Behind £200 Billion Apple?" *Daily Mail*, http://www.dailymail.co.uk/home/moslive/article-1367481/Apples-Jonathan-Ive-How-did-British-polytechnic-graduate-design-genius.html, last modified, March 19, 2013.

15. Clive Grinyer, History, http://www.clivegrinyer.com/history.html.

16. Luke Dormehl, *The Apple Revolution: Steve Jobs, the Counter Culture and How the Crazy Ones Took Over the World* (Random House, 2012), Kindle edition.

17. Interview with Clive Gryiner, January 2013.

18. Interview with Peter Phillips, January 2013.

19. Interview with Phil Gray, January 2013.

20. Dormehl, *The Apple Revolution*, Kindle edition.

21. Ibid.

22. Peter Burrows, "Who Is Jonathan Ive?" *Businessweek*, http://www.businessweek.com/stories/2006-09-24/who-is-jonathan-ive, Septermber 26, 2006.

23. Dormehl, *The Apple Revolution*, Kindle edition.

24. Jonathan Ive, Travel and attachment report, http://www.thersa.org/about-us/history-and-archive/archive/archive-search/archive/r31382, 1987–1988, 1988–1989.

25. The Royal Society for the Encouragement of Arts, Manufactures and Commerce, History, http://www.thersa.org/about-us/history-and-archive.

26. Interview with Craig Mounsey, March 2013.

27. Interview with Barry Weaver, January 2013.

28. Ibid.

29. Interview with David Tonge, January 2013.

30. The Design Council Collection, The Design Council/The Manchester Metropolitan University, Design Council, Design Centre, Haymarket,

London. Young Designers Centre Exhibition 1989. Radio hearing aid designed by Jonathan Ive of Newcastle Polytechnic. http://vads.ac.uk/large.php?uid=114262&sos=0.

31. Burrows, "Who is Jonathan Ive?"

32. Dormehl, *The Apple Revolution*, Kindle edition.

33. Melanie Andrews, "Jonathan Ive & the RSA's Student Design Awards" RSA's Design and Society blog, http://www.rsablogs.org.uk/category/design-society/page/3/, May 25, 2012.

34. London Design Museum, interview with Jonathan Ive, http://design museum.org/design/jonathan-ive, last modified 2007.

35. Ibid.

CHAPTER 3

Life in London

1. Robert Brunner Facebook page, https://www.facebook.com/robertbrun nerdesigner/info.

2. Interview with Robert Brunner, March 2013.

3. Ibid.

4. Ibid.

5. Melanie Andrews, "Jonathan Ive & the RSA's Student Design Awards" RSA's Design and Society blog, http://www.rsablogs.org.uk/category/design-society/page/3/, May 25, 2012.

6. Interview and e-mails with Barrie Weaver, January 14, 2013.

7. Interview with Phil Gray, January 2013.

8. Ibid.

9. Interview and e-mails with Barrie Weaver, January 2013

10. Interview with Clive Grinyer, January 2012.

11. Ibid.

12. Documents provided by Martin Darbyshire, May 2013.

13. Ibid.

14. Luke Dormehl, *The Apple Revolution: Steve Jobs, the Counter Culture and How the Crazy Ones Took Over the World* (Random House, 2012), Kindle edition.

15. Interview with Paul Rodgers, October 2012.

16. Interview with Clive Grinyer, January 2012.

17. Dormehl, *The Apple Revolution*, Kindle edition.

18. Interview with David Tonge, January 2013.

19. Interview with Peter Phillips, January 2013.

20. Paul Kunkel, *AppleDesign*, (New York: Graphis Inc., 1997), 254.

21. Ibid.

22. Interview with Clive Grinyer, January 2013.

23. Ibid.

24. Kunkel, *AppleDesign*, 254

25. Interview with Clive Grinyer, January 2013

26. Documents provided by Martin Darbyshire, May 2013.

27. Peter Burrows, "Who is Jonathan Ive?" *Bloomberg Businessweek*, http://www.businessweek.com/stories/2006-09-24/who-is-jonathan-ive, September 26, 2006.

28. Design Museum interview, http://designmuseum.org/design/jonathan -ive.

29. Interview with Phil Gray, January 2013.

30. Interview with Robert Brunner, March 2013.

31. John Sculley on Steve Jobs, YouTube, www.youtube.com/watch?v=S_JYy_0XUe8.

32. Harry McCracken, "Newton Reconsidered," *Time*, http://techland.time .com/2012/06/01/newton-reconsidered/, June 1, 2012.

33. Kunkel, *AppleDesign*, 237–38.

34. Ibid.

35. Dormehl, *The Apple Revolution*, Kindle edition.

36. London Design Museum, interview with Jonathan Ive, http://design museum.org/design/jonathan-ive, last modified 2007.

37. Kunkel, *AppleDesign*, 236–46.

38. Interview with Robert Brunner, March 2013.

39. Dormehl, *The Apple Revolution*, Kindle edition.

40. Interview with Martin Darbyshire, May 2013.

41. Kunkel, *AppleDesign*, 254.

42. Dormehl, *The Apple Revolution*, Kindle edition.

43. Interview with Robert Brunner, March 2013.

44. Ibid.

45. Dormehl, *The Apple Revolution*, Kindle edition.

46. Ibid.

47. Interview with Robert Brunner, March 2013.

48. Kunkel, *AppleDesign*, p. 255.

49. Dormehl, *The Apple Revolution*, Kindle edition.

50. Interview with Peter Phillips, January 2013.

51. Peter Burrows, "Who Is Jonathan Ive?" *Businessweek*, originally in Radical Craft Conference, the Art Center College of Design in Pasadena, California., http://www.businessweek.com/stories/2006-09-24/who-is -jonathan-ive

52. Design Museum, http://designmuseum.org/design/jonathan-ive.

53. Ibid.

54. Interview with Peter Phillips, Spring 2013.

CHAPTER 4

Early Days at Apple

1. John Markoff, "At Home with Jonathan Ive: Making Computers Cute Enough to Wear," http://www.nytimes.com/1998/02/05/garden/at -home-with-jonathan-ive-making-computers-cute-enough-to-wear .html, published Feruary 05, 1998.

2. Paul Kunkel, *AppleDesign*, (New York: Graphis Inc., 1997), p. 81.

3. Interview with Robert Brunner, March 2013.

4. Ibid.

5. Ibid.

6. Interview with Rick English, December 2012.

7. Ibid.

8. Paul Kunkel, *AppleDesign*, 229–30.

9. Ibid.

10. College of Creative Arts, Massey University, http://creative.massey .ac.nz.

11. Kunkel, *AppleDesign*, 253.

12. Interview with Robert Brunner, March 2013.

13. Ibid.

14. Ibid.

15. Kunkel, *AppleDesign*, 253–56.

16. Ibid.

17. Ibid.

18. Ibid.

19. Ibid., 256.

20. Ibid.

21. Ibid.
22. Kunkel, *AppleDesign*, 258.
23. Interview with Rick English, December 2012.
24. Poornima Gupta and Dan Levine, "Apple Designer: iPhone Crafters Are 'Maniacal'," http://www.reuters.com/article/2012/08/01/us-apple-samsung-designer-idUSBRE87001O20120801, July 31, 2012.
25. Kunkel, *AppleDesign*, 266.
26. Ibid., 265.
27. Interview with Don Norman, September 2012.
28. Paul Kunkel, *AppleDesign*, 272.
29. Ibid., 274.
30. Ibid., 275.
31. Ibid., 272–77.
32. Interview with Clive Grinyer, January 2013.
33. Jim Carlton, *Apple: The Inside Story of Intrigue, Egomania and Business Blunders* (HarperBusiness, 1997), p. 412.
34. Kunkel, *AppleDesign*, p. 65.
35. Daniel Turner, *MIT Technology Review* 2007, http://www.technologyreview.com/Biztech/18621/, May 1, 2007.
36. Ibid.
37. Ibid.
38. Rachel Metz, "Behind Apple's Products is Longtime Designer Ive," Associated Press, http://usatoday30.usatoday.com/tech/news/story/2011-08-26/Behind-Apples-products-is-longtime-designer-Ive/50150410/1, updated 8/26/2011.
39. Isaacon, *Steve Jobs*, Kindle edition.
40. Interview with Jon Rubinstein, October 2012.

CHAPTER 5

Jobs Returns to Apple

1. Walter Isaacson, *Steve Jobs* (Simon & Schuster, 2011), Kindle edition.
2. Steve Jobs at Apple's Worldwide Developers Conference 1998, video, http://www.youtube.com/watch?v=YJGcJgpOU9w.
3. Apple 10K Annual Report 1998: http://investor.apple.com/secfiling.cfm?filingID=1047469-98-44981&CIK=320193; and Apple 10K Annual

Report 1995: http://investor.apple.com/secfiling.cfm?filingID=320193
-95-16&CIK=320193.

4. Isaacson, *Steve Jobs*, Kindle edition.

5. Ibid.

6. Rob Walker, "The Guts of a New Machine" *New York Times*, http://www
.nytimes.com/2003/11/30/magazine/the-guts-of-a-new-machine.html,
November 30, 2003.

7. Paul Kunkel, *AppleDesign* (New York, Graphis Inc., 1997) p. 21.

8. Ibid., 24.

9. Ibid., 26.

10. Isaacson, *Steve Jobs*, Kindle edition. Named after the Bob Dylan song
"Stuck inside of Mobile with the Memphis Blues Again." The group's
leader was a fan.

11. Bertrand Pellegrin, "Collectors Give '80s Postmodernist Design 2nd
Look," *San Francisco Chronicle*, http://www.sfgate.com/homeandgarden/
article/Collectors-give-80s-postmodernist-design-2nd-look-2517937
.php, January 15, 2012.

12. Andy Reinhardt, "Steve Jobs on Apple's Resurgence: Not a One-Man
Show,"*Businessweek*, http://www.businessweek.com/bwdaily/dnflash/
may1998/nf80512d.htm, May 12, 1998.

13. Bill Buxton, *Sketching User Experiences: Getting the Design Right and the
Right Design* (Morgan Kaufman, 2007) 41–42.

14. Ibid.

15. Peter Burrows, "Who is Jonathan Ive?" *Bloomberg Businessweek*, http://
www.businessweek.com/stories/2006-09-24/who-is-jonathan-ive.

16. Isaacson, *Steve Jobs*, Kindle edition.

17. Interview with Doug Satzger, January 2013.

18. Alan Deutschman, *The Second Coming of Steve Jobs* (Random House,
2001), 251.

19. Isaacson, *Steve Jobs*, Kindle edition.

20. Jennifer Tanaka, "No More Beige boxes," *Newsweek*, http://www.thedai
lybeast.com/newsweek/1998/05/18/what-inspires-apple-s-design-guru
.html, 05/18/1998.

21. Kunkel, *AppleDesign*, 280.

22. Delphine Hirasuna, "Sorry, No Beige," *Apple Media Arts*, vol. 1, no. 2:
p. 4, http://timisnice.blogspot.com/2011/02/interviewing-jonathan
-ive-delphine.html, Summer 1998.

23. Ibid.
24. Interview with Paul Dunn, July 2013.
25. Mark Prigg, "Sir Jonathan Ive: The iMan Cometh," *London Evening Standard*, http://www.standard.co.uk/lifestyle/london-life/sir-jonathan-ive-the-iman-cometh-7562170.html, March 12, 2012.
26. Interview with Marj Andresen, December 2012.
27. Interview with Roy Askeland, July 2013.
28. Interview with Paul Dunn, July 2013.
29. Isaacson, *Steve Jobs*, Kindle edition.
30. Dike Blair, "Bondi Blue," Interview with Jonathan Ive for *Purple #2*, Winter 98/99, 268–75.
31. Isaacson, *Steve Jobs*, Kindle edition.
32. Benj Edwards, "The Forgotten eMate 300—15 Years Later," originally in *Macweek*, December 21, 2012.
33. Interview with Doug Satzger, January 2013.
34. Isaacson, *Steve Jobs*, Kindle edition.
35. Apple brochure from 1977, noted in Walter Isaacson, *Steve Jobs*.
36. David Kirkpatrick, reporter associate Tyler Maroney, "The Second Coming of Apple Through a Magical Fusion of Man—Steve Jobs—and Company, Apple Is Becoming Itself Again: The Little Anticompany That Could," *Fortune*, http://money.cnn.com/magazines/fortune/fortune_archive/1998/11/09/250834/, November 9, 1988.
37. Blair, "Bondi Blue."
38. Isaacson, *Steve Jobs*, Kindle edition.
39. Ibid.
40. Interview with Don Norman, September 2012.
41. Leander Kahney, "Interview: The Man Who Named the iMac and Wrote Think Different," *Cult of Mac*, http://www.cultofmac.com/20172/20172/, November 3, 2009.
42. Ibid.
43. Ibid.
44. Ibid.
45. Ibid.
46. Interview with Amir Homayounfar, April 2013.
47. Isaacson, *Steve Jobs*, Kindle edition.
48. Ibid.
49. Jodi Mardesich, "Macintosh Power Play $1,299 PC: A Combination of

Techno-Lust and Fashion Envy; It'll Be Available in 90 Days," *San Jose Mercury News*, May 7, 1998.

50. Steve Jobs, Apple Special Event, introduction of the iMac, May 6, 1998. http://www.youtube.com/watch?v=oxwmF0OJ0vg.

51. Interview with Doug Satzger, January 2013.

52. Hiawatha Bray, "Thinking Too Different," *Boston Globe*, May 4, 1998

53. Matt Beer, "New Unit Built with Users, Not Engineers, in Mind," *Vancouver Sun*, August 13, 1998.

54. "Will iMac Ripen Business for Apple?" *Associated Press*, published on Cnn.com, http://www.cnn.com/TECH/computing/9808/15/imac/, August 15, 1998.

55. "Apple Computer's Futuristic New iMac Goes on Sale," Associated Press, http://chronicle.augusta.com/stories/1998/08/15/tec_236131.shtml, August 15, 1998.

56. Jon Fortt, "New iMac Friendlier, but Apple Falls Short," *San Jose Mercury News*, January 14, 2002.

57. Theresa Howard, "See-Through Stuff Sells Big: iMac Inspires Clear Cases for Other Gadgets," *USA TODAY*, December 26, 2000.

58. Interview with Penny Sparke, September 2012.

59. Howard, "See-Through Stuff."

60. CNET News.com staff, "Gates Takes a Swipe at iMac," CNET, http://news.cnet.com/Gates-takes-a-swipe-at-iMac/2100-1001_3-229037.html, July 26, 1999.

61. James Culham, "Forever Young: From Cars to Computers to Furniture, the Current Colourful, Playful, Almost Toy-like Design Esthetic Owes More to the Playhouse Than the Bauhaus," *Vancouver Sun*, February 10, 2001.

62. Isaacson, *Steve Jobs*, Kindle edition.

63. Hirasuna, "Sorry, no beige."

CHAPTER 6

A String of Hits

1. Lev Grossman, "How Apple Does It," *Time*, http://www.time.com/time/magazine/article/0,9171,1118384,00.html, October 16, 2005.

2. Interview with Doug Satzger, January 2013.

3. Walter Isaacson, *Steve Jobs* (Simon & Schuster, 2011), Kindle edition.

4. Interview with a former Apple executive, December 2012.
5. Phil Schiller testimony during *Apple v. Samsung* trial, trial transcript online at Groklaw (but behind paywall).
6. Interview with a former Apple executive, December 2012.
7. Interview with Sally Grisedale, February 2013.
8. Grossman, "How Apple Does It."
9. Neil Mcintosh, "Jobs Unveils the G4 Super Mac," *Guardian*, http://www.guardian.co.uk/technology/1999/sep/02/onlinesupplement1, September 1, 1999.
10. Isaacson, *Steve Jobs*, Kindle edition.
11. Interview with Jon Rubinstein, October 2012.
12. Mark Prigg, "Sir Jonathan Ive: The iMan Cometh," *London Evening Standard*, http://www.standard.co.uk/lifestyle/london-life/sir-jonathan-ive-the-iman-cometh-7562170.html.
13. Jonathan Ive in Apple iBook G3 Introduction, 2007, video, http://www.youtube.com/watch?v=_X9PWjUD9gU
14. Henry Norr, iBook Looks Less Different: This Time, Enternal Features Distinquish Apple's Noebook, *San Francisco Chronicle*, http://www.sfgate.com/business/article/REVIEW-iBook-looks-less-different-This-time-2920054.php, May 17, 2001.
15. Steve Gillmor, "Off the Record," *Infoworld*, http://www.infoworld.com/d/developer-world/record-937 October 21, 2002.
16. Interview with Jon Rubinstein, October 2012.
17. Jonathan Ive, "Celebrating 25 Years of Design" Design Museum 2007, http://designmuseum.org/design/jonathan-ive.
18. Christopher Stringer testimony during *Apple v. Samsung* trial, trial transcript online at Groklaw (but behind paywall).
19. Ibid.
20. Simon Jarry, "2001MW Expo: Titanium G4 PowerBook stunner," Macworld UK, http://www.macworld.co.uk/mac/news/?newsid=2323, January 10, 2001.
21. Jonathan Ive in Apple iBook G3 Introduction, 2007.
22. Ibid.
23. John Siracusa, "G4 Cube & Cinema Display," http://archive.arstechnica.com/reviews/4q00/g4cube_cd/g4-cube-3.html, October 2000.
24. Andrew Gore, "The Cube," Macworld.com, http://www.macworld.com/article/1015641/buzzthe_cube.html, October 1, 2000.

25. John Siracusa, "G4 Cube & Cinema Display."
26. Apple.com press releases, "Apple to Report Disappointing First Quarter Results," http://www.apple.com/pr/library/2000/12/05Apple-To-Report-Disappointing-First-Quarter-Results.html, December 5, 2000.
27. Brad Gibson, "Macworld: Numbers Tell the Story for Apple Sales," *PC World*, http://www.macworld.com/article/1021753/apple.html, January 19, 2001.

CHAPTER 7

The Design Studio Behind the Iron Curtain

1. Charles Piller, "Apple Finds Its Design Footing Again with iMac" *LA Times*, http://articles.latimes.com/1998/jun/08/business/fi-57794, June 8, 1998.
2. Marcus Fairs, "Jonathan Ive," *ICON*, http://www.iconeye.com/read-previous-issues/icon-004-|-july/august-2003/jonathan-ive-|-icon-004-|-july/august-2003, July/August 2003.
3. Lev Grossman, "How Apple Does It," *Time*, http://www.time.com/time/magazine/article/0,9171,1118384,00.html, October 16, 2005.
4. Interview with a former Apple engineer, June 2013.
5. Interview with Jon Rubinstein, October 2012.
6. Interview with Doug Satzger, January 2013.
7. Christopher Stringer testimony, *Apple v. Samsung* trial, San Jose Federal Courthouse, July 2012.
8. Ibid.
9. Ibid.
10. Interview with Gautam Baksi, June 2013.

CHAPTER 8

Design of the iPod

1. Steven Levy, *The Perfect Thing: How the iPod Shuffles Commerce, Culture, and Coolness* (Simon & Schuster, 2006), 36.
2. Ibid., 38.
3. Ibid., 133.
4. Sheryl Garratt, "Jonathan Ive: Inventor of the decade," *The Guartdian*,

http://www.guardian.co.uk/music/2009/nov/29/ipod-jonathan-ive
-designer, November 28, 2009.

5. Leander Kahney, "Straight Dope on the IPod's Birth," http://www.wired
    .com/gadgets/mac/commentary/cultofmac/2006/10/71956, October
    26, 2006.

6. Interview with Tim Wasko, April 2013.

7. Rob Walker, "The Guts of a New Machine," *New York Times Magazine*,
    http://www.nytimes.com/2003/11/30/magazine/the-guts-of-a-new
    -machine.html?pagewanted=all&src=pm, November 30, 2003.

8. Jonathan Ive, "iPod—2001 and 2002," Design Museum Online Exhi-
    bition,http://designmuseum.org/exhibitions/online/jonathan-ive-on
    -apple/ipod-emac

9. Interview with Chris Lefteri, October 2012.

10. Levy, *The Perfect Thing*, 78.

11. Ibid., 99–100.

12. Jonathan Ive in Apple Original iPod Introduction, 2006, video, http://
    www.youtube.com/watch?v=TSqNHGJw2qI

13. Levy, *The Perfect Thing*, 50.

14. Macslah Forum post, found on "Apple's 'breakthrough' iPod," Brad
    King and Farhad Manjoo, Wired.com, http://www.wired.com/gadgets/
    miscellaneous/news/2001/10/47805, October 23, 2001.

15. Apple—"Introducing the First iPod," http://www.youtube.com/
    watch?v=BCYhrt_PF7Q

16. Levy, *The Perfect Thing*, 51.

17. Johnny Davis, "Ten Years of the iPod," *The Guardian* (UK), http://www
    .guardian.co.uk/technology/2011/mar/18/death-ipod-apple-music,
    March 17, 2011.

CHAPTER 9

Manufacturing, Materials and Other Matters

1. Walter Isaacson, *Steve Jobs* (Simon & Schuster, 2011), Kindle edition.

2. Ibid.

3. Garry Barker, "The i of the Beholder," interview with Jonathan Ive,
    Sydney Morning Herald, http://www.smh.com.au/articles/2002/06/19/
    1023864451267.html, June 19, 2002.

4. Henry Norr, "Apple's New iMac: Team Develops Unique Ideas," *San Francisco Chronicle*, January 8, 2002.

5. Ibid.

6. Barker, "The i of the Beholder."

7. Isaacson, *Steve Jobs*, Kindle edition.

8. Apple iMac G4 , 2011, video, http://www.youtube.com/watch?v=0Ky_vx FBeJ8

9. "Apple Takes a Bold New Byte at iMac," *New Zealand Herald*, http://www .nzherald.co.nz/technology/news/article.cfm?c_id=5&objec-tid=787149, January 21, 2002.

10. Email from Ken Segall, April 2013.

11. Interview with Dennis Boyle, October 2012.

12. Steven Levy, "The New iPod" *Newsweek*, http://www.thedailybeast .com/newsweek/2004/07/25/the-new-ipod.html, July 25, 2004.

13. Steven Levy, *The Perfect Thing: How the iPod Shuffles Commerce, Culture, and Coolness* (Simon & Schuster, 2006) 102.

14. Jonathan Ive in conversation with Dylan Jones, editor of British *GQ*, following his award of honorary doctor at the University of the Arts London, © Nick Carson 2006. First published in issue 5 of *TEN4*, http:// ncarson.wordpress.com/2006/12/12/jonathan-ive/, November 16, 2006.

15. IDEA, www.idsa.org/award

16. *ICON* 004, http://www.iconeye.com/read-previous-issues/icon-004| -july/august-2003/jonathan-ive-|-icon-004-|-july/august-2003, July/ August 2003.

17. Neil Mcintosh, "Return of the Mac," http://www.guardian.co.uk/ artanddesign/2003/jun/04/artsfeatures.shopping, June 3, 2003.

18. Ibid.

19. Garry Barker, "Hey Mr. Tangerine Man," *Sydney Morning Herald*, http://www .smh.com.au/articles/2003/06/11/1055220639850.html, June 12, 2003.

20. Nathalie Atkinson, "That New White Magic," *Saturday Post*, Canada, http://www.nationalpost.com/search/site/story.asp?id=1378CAFA -0509-4389-8B7E-4333915AF45A, August 2, 2003.

21. Larry Elliott, "Better Design Requires Better Product," http://www .guardian.co.uk/business/2005/nov/21/politics.economicpolicy, No-vember 20, 2005.

22. Ibid.

23. Jonny Evans, "Apple Design Chief Jonathan Ive Collects CBE,"Macworld, http://www.macworld.co.uk/mac/news/?newsid=16510, November 17, 2006.

24. Marcus Fairs, *ICON*, http://www.iconeye.com/read-previous-issues/icon-004-|-july/august-2003/jonathan-ive-|-icon-004-|-july/august-2003, July /August 2003.

25. Dick Powell, "At the Core of Apple," Innovate, issue 6, Summer 2009, http://www.innovation.rca.ac.uk/cms/files/Innovate6.pdf

26. Phil Schiller, *Apple v. Samsung* trial testimony.

27. Isaacson, *Steve Jobs*, Kindle Edition.

28. Apple Press info, "Tim Cook Named COO of Apple," http://www.apple.com/pr/library/2005/10/14Tim-Cook-Named-COO-of-Apple.html, October 14, 2005.

29. Interview with Jon Rubinstein, October 2012

30. Leander Kahney, *Inside Steve's Brain*, expanded edition (Portfolio, 2009), 96.

31. Joel West, "Apple Computer: The iCEO Seizes the Internet," http://www.scribd.com/doc/60250577/APPLE-Business, October 20, 2002.

32. Adam Lashinsky, "Tim Cook: The Genius Behind Steve," *Fortune*, http://money.cnn.com/2011/08/24/technology/cook_apple.fortune/index.htm, August 24, 2011.

33. Interview with Doug Satzger, January 2013.

34. Walter Isaacson, *Steve Jobs*, Kindle edition

35. Ibid.

CHAPTER 10

The iPhone

1. Walter Isaacson, *Steve Jobs* (Simon & Schuster) Kindle Edition.

2. Ibid.

3. Ibid.

4. John Paczkowski, "Apple CEO Steve Jobs Live at D8," http://allthingsd.com/20100601/steve-jobs-session/, June 1, 2010.

5. Scott Forstall, *Apple v. Samsung* trial testimony.

6. Ibid.

7. Kevin Rose, "Matt Rogers: Founder of Nest Labs interview," *Foundation* 21, 2012, video, http://www.youtube.com/watch?v=HegU77X6I2A

8. "On the verge," *The Verge*, April 29, 2012, video http://www.theverge
.com/2012/4/30/2987892/on-the-verge-episode-005-tony-fadell-and
-chris-grant.

9. Isaacson, *Steve Jobs*, Kindle Edition.

10. Scott Forstall testimony at *Apple v. Samsung* trial.

11. *Apple v. Samsung* trial, deposition of Jonathan Ive,

12. Ibid.

13. Ibid.

14. Ibid.

15. Charles Duhigg and Keith Bradsher, "How the U.S. Lost Out on iPhone
Work," *New York Times*, http://www.nytimes.com/2012/01/22/business/
apple-america-and-a-squeezed-middle-class.html, January 21, 2012.

16. Bryan Gardiner, "Glass works: How Corning Created the Ultrathin,
Ultrastrong Material of the Future," Wired.com, http://www.wired.
com/wiredscience/2012/09/ff-corning-gorilla-glass/all/, September 24,
2012.

17. Isaacson, *Steve Jobs*, Kindle edition.

18. Fred Vogelstein, "The Untold Story: How the iPhone Blew Up the Wireless Industry," Wired.com, http://www.wired.com/gadgets/wireless/
magazine/16-02/ff_iphone, January 1, 2008.

19. Katherine Rushton, "Apple Design Chief Sir Jonathan Ive: iPhone was
'Nearly Axed,'" http://www.telegraph.co.uk/technology/apple/9440639/
Apple-design-chief-Sir-Jonathan-Ive-iPhone-was-nearly-axed.html,
July 31, 2012.

20. Vogelstein, "The Untold Story."

21. *Apple v. Samsung* trial, testimony of Christopher Stringer.

22. Janko Roettgers, "Alan Kay: With the Tablet, Apple Will Rule the
World," Gigaom.com, http://gigaom.com/2010/01/26/alan-kay-with-
the-tablet-apple-will-rule-the-world/.

CHAPTER 11

The iPad

1. Walter Isaacson, *Steve Jobs* (Simon & Schuster, 2011), Kindle edition.

2. Brian Heater, "Steve Jobs Shows No Love for Netbooks," http://www.pc
mag.com/article2/0,2817,2358514,00.asp, January 28, 2010.

3. Isaacson, *Steve Jobs*, Kindle edition.

4. Ibid.

5. *Apple v. Samsung* trial, testimony of Christopher Stringer.

6. Charles Arthur, "Netbooks Plummet While Tablets and Smartphones Soar, says Canalys," The *Guardian*, http://www.guardian.co.uk/technol ogy/blog/2012/feb/03/netbooks-pc-canalys-tablet, February 3, 2012.

7. Apple Press info, "Mark Papermaster Joins Apple as Senior Vice President of Devices Hardware Engineering," http://www.apple.com/pr/ library/2008/11/04Mark-Papermaster-Joins-Apple-as-Senior-Vice -President-of-Devices-Hardware-Engineering.html, November 4, 2008.

8. Fadell declined to comment.

9. Jonathan Ive in Apple iPad 2 official video 2011, video, http://www .youtube.com/watch?v=fjlvmbJEUmk, March 2011.

10. Ibid.

11. David Pogue, "This Year, Gift Ideas in Triplicate," *New York Times*, http://www.nytimes.com/2012/11/01/technology/personaltech/ presenting-the-nook-hd-ipad-mini-and-windows-phone-8-review .html, October 30, 2012.

12. Gartner Inc., "Gartner Says Worldwide PC, Tablet and Mobile Phone Shipments to Grow 5.9 Percent in 2013 as Anytime-Anywhere Computing Drives Buyer Behavior," http://www.gartner.com/news room/id/2525515, June 24, 2013.

CHAPTER 12

Unibody Everywhere

1. Apple special event video, Oct 14: Apple Notebook Event 2008, New Way to Build 2-/6, 2008, video, http://www.youtube.com/watch?v=7JL jldgjuKI.

2. Ibid.

3. Interview with Doug Satzger, January 2013.

4. Apple special event video, Oct 14.

5. Interview with Chris Lefteri, October 2012.

6. Interview with a former Apple engineer, June 2013.

7. Personal interview, June 2013.

8. Interview with Dennis Boyle, October 2012.

9. Interview with a former Apple engineer, June 2013.

10. Horace Dediu, "How Much Do Apple's Factories Cost?" http://www

.asymco.com/2011/10/16/how-much-do-apples-factories-cost/ October 16, 2011.

11. Greenpeace, "Guide to Greener Electronics 18," http://www.greenpeace. org/new-zealand/en/Guide-to-Greener-Electronics/18th-Edition/AP PLE/, November 2012.

12. Interview with Kyle Wiens, June 2013.

13. Interview with Doug Satzger, January 2013.

CHAPTER 13

Apple's MVP

1. Walter Isaacson, *Steve Jobs* (Simon & Schuster, 2011), Kindle edition.

2. Ibid.

3. Ibid.

4. Jemima Kiss, "Apple's Worst Nightmare: Is Jonathan Ive to Leave?" http://www.theguardian.com/technology/pda/2011/feb/28/apple -jonathan-ive, February 28, 2011.

5. Maurice Chittenden and Sean O'Driscoll, "I Created the iPad and iClaim my £18m," http://www.thesundaytimes.co.uk/sto/news/uk_news/ Tech/article563855.ece, February 27, 2011.

6. Martha Mendoza, "Apple Designer as Approachable as His iMac," Associated Press, April 8 1999.

7. Shane Richmond, "Jonathan Ive interview: Apple's Design Genius is British to the Core," *Telegraph*, http://www.telegraph.co.uk/technology/ apple/9283486/Jonathan-Ive-interview-Apples-design-genius-is -British-to-the-core.html, May 23, 2012.

8. Ibid.

9. Interview with Phil Gray, January 2013.

10. Apple WWDC 2010—iPhone 4 Introduction, 2010, video, http://www .youtube.com/watch?v=z__jxoczNWc.

11. Apple Press info, "Apple Announces Changes to Increase Collaboration Across Hardware, Software & Services," http://www.apple.com/pr/ library/2012/10/29Apple-Announces-Changes-to-Increase -Collaboration-Across-Hardware-Software-Services.html, October 29, 2012.

12. Mark Gurman, "Tim Cook Emails Employees, Thanks Scott Forstall, Says Bob Mansfield to Stay On for Two Years," http://9to5mac.com/2012/10/29/tim-cook-emails-employees-thanks-scott-forstall-says-bob-mansfield-to-stay-on-for-two-years/, October 29, 2012.

13. Nick Wingfield and Nick Bilton, "Apple Shake-Up Could Lead to Design Shift," *New York Times*, http://www.nytimes.com/2012/11/01/technology/apple-shake-up-could-mean-end-to-real-world-images-in-software.html, October 31, 2012.

14. Shane Richmond, "Jonathan Ive Interview: Simplicity Isn't Simple," *Telegraph*, http://www.telegraph.co.uk/technology/apple/9283706/Jonathan-Ive-interview-simplicity-isnt-simple.html, May 23, 2012.

15. Apple 2013 Worldwide Developers Conference, keynote video: http://www.youtube.com/watch?v=qzUH9PJA1Ro, June 10, 2013.

16. Interview with Sally Grisedale, February 2013.

17. Interview with Larry Barbera, June 2013.

18. Alex Schleifer, "The Age of the User Interface," http://saydaily.com/2013/02/design-really-is-everything-now.html, February 15, 2013.

19. Josh Tyrangiel, "Tim Cook's Freshman Year: The Apple CEO Speaks," *Bloomberg Businessweek*, http://www.businessweek.com/articles/2012-12-06/tim-cooks-freshman-year-the-apple-ceo-speaks, December 6, 2012.

20. Isaacson, *Steve Jobs*, Kindle edition.

21. Ibid.

22. Richmond, "Jonathan Ive Interview: Simplicity Isn't Simple."

23. Katherine Rushton, "Apple Design Chief: 'Our Goal Isn't to Make Money,'" *Telegraph*, http://www.telegraph.co.uk/technology/apple/9438662/Apple-design-chief-Our-goal-isnt-to-make-money.html, July 30, 2012.

24. Katherine Rushton, "Apple Design Chief Sir Jonathan Ive: iPhone Was 'Nearly Axed,'" *Telegraph*, http://www.telegraph.co.uk/finance/newsbysector/mediatechnologyandtelecoms/9440639/Apple-design-chief-Sir-Jonathan-Ive-iPhone-was-nearly-axed.html, July 31, 2012.

25. Interview with Clive Grinyer, January 2013.

26. Ibid.

27. Jonathan Ive, speaking in Objectified documentary, 2009.

28. London Design Museum, interview with Jonathan Ive, http://designmuseum.org/design/jonathan-ive, last modified 2007.

29. Mark Prigg, "Sir Jonathan Ive: Knighted for Services to Ideas and Innovation," *The Independent*, http://www.independent.co.uk/news/people/profiles/sir-jonathan-ive-knighted-for-services-to-ideas-and-innovation-7563373.html, March 13, 2012.
30. Interview with Andrew Hargadon, October 2012.
31. Interview with Alex Milton, October 2012.

INDEX

PHOTO CREDITS

Page 1: NTI

Page 2: Weaver Design London UK

Page 3, top: Weaver Design London UK

Page 3, middle and bottom: © Jim Abeles. Photo by Jonathan Zufi

Page 4: Hartmut Esslinger / frog design

Page 5: ©Rick English Pictures

Page 6, top: ©Rick English Pictures

Page 6, bottom: Associated Press / Susan Ragan

Page 7: Copyright LANCE IVERSEN/San Francisco Chronicle/Corbis

Page 8, top: Copyright ©2013 Ian Larkin

Page 8, middle: photography by Kevin Coté, all rights reserved

Page 8, bottom: Copyright 2008 Dave Lawrence. Photo courtesy of Dave Lawrence (davelawrence8 on Flickr)."

Page 9, top and middle: © Jim Abeles

Page 9, bottom: Aaron Payne

Page 10, top: "Photo of a Power Mac G5 1.8 Dual, open case. Photo by User:Grm_wnr. Reproduced here under Creative Commons Attribution 2.0 Generic License."

Page 10, middle: Anonymous

Page 10, bottom: © Jim Abeles

Page 11: © Jim Abeles

Page 12, top: Associated Press/Paul Sakuma

Page 12, bottom: Josh Lowensohn/CNET

Page 13: Associated Press/Rex Features

Page 14: Associated Press/Paul Sakuma

Page 15: Kyle Wiens/iFixit

Page 16: Noah DaCosta, www.noahdacosta.com